Prime Discoveries

April 9, 2026

Contents

Contents

What is *Prime Discoveries*?

Prime Discoveries is a discovery-driven mathematics book. It approaches prime numbers not as a finished academic subject, but as a living structure still waiting to be uncovered. It presents independent discoveries through a new mathematical language and a new style of exploration.

Built through what we call the *Virgin Mindset*—starting from zero, reducing inherited bias, and reformulating the subject from the ground up—the book develops the *Number–Wave Framework* to describe primes through waves, spectra, and shadows, and to turn hidden numerical behavior into computable structures and counting formulas.

This book is not written as a conventional textbook, and it is not intended to be a survey of the literature. It is a book of formulation, exploration, and mathematical construction. Its path moves from intuition and first principles toward exact prime and twin-prime counting, asymptotic patterns, and broader structural ideas about the world of numbers.

At the same time, *Prime Discoveries* is evolutionary. It may change substantially as the framework grows, as definitions become sharper, and as formulas are refined, corrected, extended, or even replaced. Some ideas in this book may appear in forms that the reader has never seen elsewhere. That is natural in a discovery book. New language, new models, and new formulas are part of the exploratory process.

For that reason, the reader should not treat every statement in this book as an established mathematical fact unless it is explicitly presented as a proved result. The purpose of this work is different. Its purpose is to open space for thought, formulation, testing, and further development. It aims to present mathematics in a simple and enjoyable form while still demanding deep thought from the reader.

This book also does not claim competition over priority or first publication. Some of its central ideas may already exist elsewhere in other languages, other frameworks, or other formulations. Mathematics is larger than any one presentation, and similar structures may be discovered independently by different minds.

Thinkers, mathematicians, teachers, and researchers are warmly invited to study, test, expand, refine, and build upon the frameworks, models, and formulas presented here. That is part of the spirit of this work. However, the book itself remains an authored work. Its ideas may inspire new work, but the text, structure, and presentation should be respected as original creative authorship and should not be reproduced as a substitute for the book itself.

In that sense, *Prime Discoveries* is both an invitation and a challenge: an invitation to think freely, and a challenge to examine unfamiliar ideas with care, precision, and intellectual courage.

Chapter 1

Atomic Model of Numbers

"It is not knowledge, but the act of learning, not possession but the act of getting there, which grants the greatest enjoyment."
— *Carl Friedrich Gauss*

Figure 1.1: Johann Carl Friedrich Gauss (1777 – 1855)

1.1 Preface

1.1.1 Introduction

The way we perceive numbers shapes our understanding of mathematics. The *Atomic Model of Numbers* is a framework that draws an analogy with Bohr's atomic model: it envisions each natural number n as an atom A_n, with a central nucleus and particles called *Primeons* orbiting in fixed circular paths. This viewpoint provides a visual and conceptual language for describing the structure of numbers, especially the mechanism that distinguishes primes from composites.

In this chapter, we introduce the Atomic Model of Numbers at a high level. In future chapters, after establishing the necessary foundations in basic mathematical tools, we will formalize the numerical atom and develop a framework that helps us understand prime numbers more deeply.

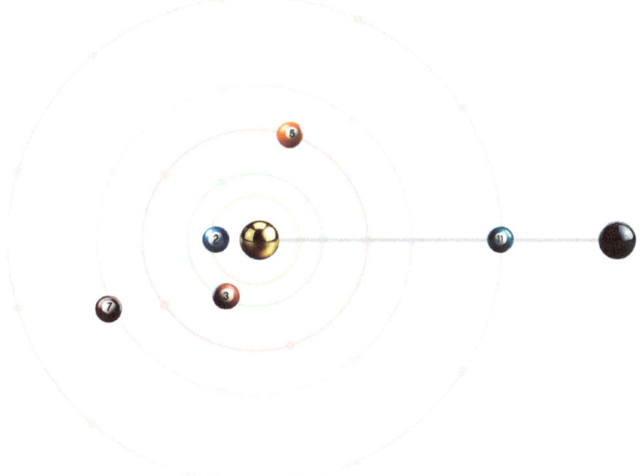

Figure 1.2: (A_{11}) — Atomic Model of the Number 11

This chapter is dedicated to **Carl Friedrich Gauss** and **Niels Bohr**: Gauss for shaping the modern language of arithmetic and congruences, and Bohr for the atomic picture that inspired the visual form of this model.

Carl Friedrich Gauss

Carl Friedrich Gauss (1777–1855) is a central figure in the history of mathematics. His work unified and elevated number theory, and his systematic treatment of congruences provided a clean language for reasoning about divisibility, remainders, and periodic structure. Because remainders and periodicity are the core mechanics behind the Atomic Model of Numbers, Gauss naturally belongs at the beginning of this journey.

Niels Bohr

> "How wonderful that we have met with a paradox. Now we have some hope of making progress."
>
> *— Niels Bohr*

Figure 1.3: Niels Bohr (1885–1962)

Niels Bohr stands as one of the central figures of modern physics, whose contributions profoundly transformed our understanding of the atom. As the architect of the Bohr model, he clarified the structure of atomic nuclei and electron orbits, laying key foundations for quantum mechanics and atomic theory. Beyond his theoretical achievements, he helped foster international scientific collaboration, and his Institute of Theoretical Physics in Copenhagen became a beacon of intellectual excellence for generations of scientists. Bohr's legacy endures not only through his scientific contributions but also through his commitment to the open exchange of ideas.

1.2 Atomic Model of Numbers

1.2.1 Analogy with Bohr's Atomic Model

In Bohr's atomic model, an atom consists of a central nucleus surrounded by electrons traveling in fixed circular orbits, with each orbit corresponding to a specific energy level. Similarly, in the Atomic Model of Numbers, each number n is represented as an atom A_n with a central nucleus. Surrounding this nucleus are particles called *Primeons*. Each Primeon has a specific energy level and moves in a fixed circular orbit whose circumference matches that energy level. This model provides a visual and conceptual way to understand the properties of numbers, particularly prime numbers.

Let us continue with an example. The Atomic Model of the number 9 has a nucleus (depicted in gold) and several Primeons orbiting around it.

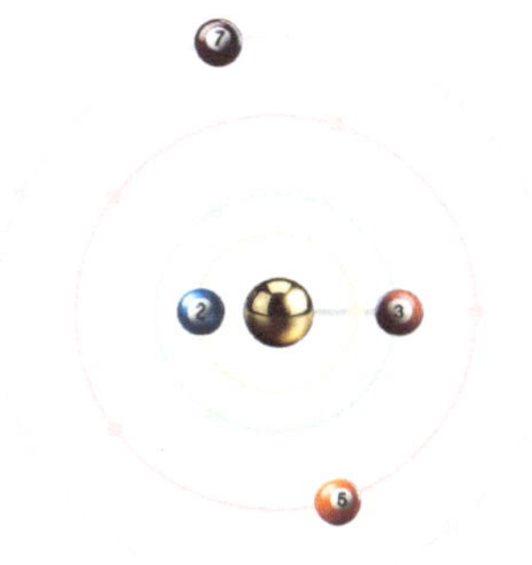

Figure 1.4: (A_9) — Atomic Model of the Number 9

From now on, when we refer to an *atom*, we mean an atom in the Atomic Model of Numbers, not in physics. Also, *number* refers to a natural number, and *Primeons* are the particles that encode prime information in this model. We also shorten *prime number* and *composite number* to *prime* and *composite*, respectively. These definitions and terms are specific to this journey and are used consistently throughout the entire work.

1.3 Structure of Numerical Atoms

All numerical atoms have a nucleus at the center, and all atoms except A_1 have several *Primeons* orbiting around it. Therefore, in general, each atom has two key components: a nucleus and a collection of Primeons.

An atom with just a single nucleus and no Primeons is the atom of number 1, denoted A_1.

Figure 1.5: (A_1) — Atom of 1

A_1 is the only atom that has no Primeons.

Now, let us learn how larger atoms emerge through interactions between nuclei.

1.4   Generation of Numerical Atoms

In this section, we explore how numerical atoms are generated over time in the Atomic Model of Numbers.

We have seen that A_1 consists of just one nucleus. To generate larger atoms, interactions between nuclei are necessary. Consider placing another A_1 close to the original A_1 to initiate interaction. We show the nucleus of the second atom in black.

Figure 1.6: Two A_1 Atoms Meet

There are fundamental rules governing how the Atomic Model of Numbers evolves at each second t. We define them next.

1.4.1   Fundamental Rules for Generating Numerical Atoms

To generate larger atoms sequentially, we consider the interaction between two nuclei (or two A_1 atoms). At each second t, these nuclei emit energies toward each other along a straight line. We call this line the *critical line*. The rules governing their interaction are as follows:

- **Time 1 — Critical Line:** At time $t = 1$, two A_1 atoms come close and establish a connection through a direct line defined as the critical line. All Primeons will be generated on this line.

- **Energy Emission:** From time $t = 2$ onward ($t \geq 2$), at each second the two nuclei emit energies toward each other along the critical line.

- **Energy Collision and Generation of Primeons:** If, at time t, the critical line is clear (unobstructed) so that no existing Primeon blocks the energies, the energies collide, resulting in the generation of a new Primeon.

- **Energy Level and Orbits:** The energy level of the newly generated Primeon is equal to t. It begins orbiting around the nucleus in a circular path with a circumference of t units.

- **Energy Absorption:** If, at time t, the critical line between the nuclei is obstructed by any existing Primeon, the emitted energies are absorbed by the obstructing Primeon.

- **Speed of Primeons:** The speed of all Primeons remains uniform at $1\,\mathrm{m/s}$.

1.4.2   Illustration of the Generation of Numerical Atoms

We now follow the generation process step by step through illustrations.

Second 2 ($t = 2$)

At $t = 2$, both nuclei emit their first energies toward each other, and the energies collide without any obstruction.

Figure 1.7: Generation of the First Primeon with Energy Level 2 at $t = 2$

This collision generates the first (and most important) Primeon, denoted pr_1, at time $t = 2$. The energy level of pr_1 is 2, so the Primeon moves into a fixed orbit with circumference 2 units around the nucleus.

Figure 1.8: Atomic Model of Number 2 (A_2)

To distinguish Primeons, their energy levels are written on them.

Second 3 ($t = 3$)

At $t = 3$, the emitted energies collide again without obstruction, generating the second Primeon pr_2 with energy level 3.

Figure 1.9: Generation of the Second Primeon at $t = 3$

pr_2 starts orbiting in its orbit of circumference 3 units.

Figure 1.10: Atomic Model of Number 3 (A_3)

Second 4 ($t = 4$)

At $t = 4$, the critical line between the nuclei is obstructed by pr_1. According to the fundamental rules, the energies emitted by the nuclei are absorbed by pr_1, so no new Primeon is generated at $t = 4$.

Figure 1.11: Attempted Collision at $t = 4$ Is Blocked by the First Primeon pr_1

- Since pr_1 is placed on the critical line, the energy level of pr_1, which is 2, is a prime factor of $t = 4$.

Second 5 ($t = 5$)

At $t = 5$, the critical line is clear again, and the energies collide to generate the third Primeon pr_3 with energy level 5.

Figure 1.12: Generation of the Third Primeon at $t = 5$

pr_3 orbits with circumference 5 units.

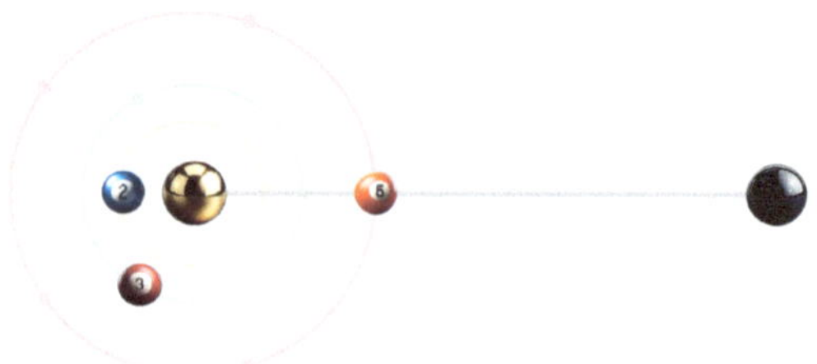

Figure 1.13: Atomic Model of Number 5 (A_5)

The generation of a Primeon at time t indicates that A_t is a prime atom, meaning t is prime. Therefore, 5 is prime.

Second 6 ($t = 6$)

At $t = 6$, the critical line is obstructed by both pr_1 and pr_2, indicating that 6 is composite with prime factors 2 and 3. The energies emitted by the nuclei are absorbed by pr_1 and pr_2.

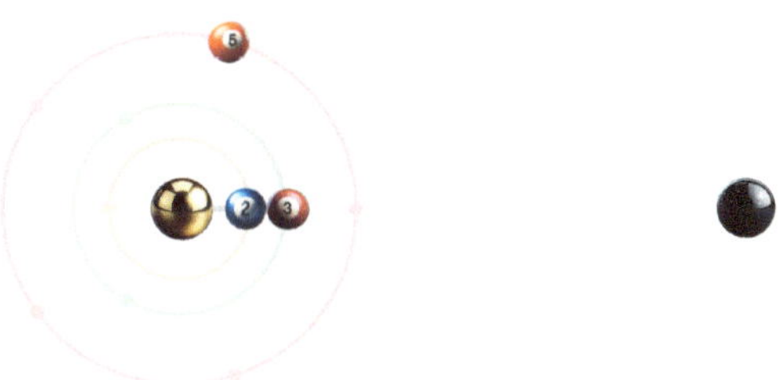

Figure 1.14: Atomic Model at $t = 6$ (A_6)

The energy levels of Primeons placed on the critical line at time t are distinct prime factors of t.

Second 7 ($t = 7$)

At $t = 7$, the energies collide unobstructed, creating the fourth Primeon pr_4 with energy level 7.

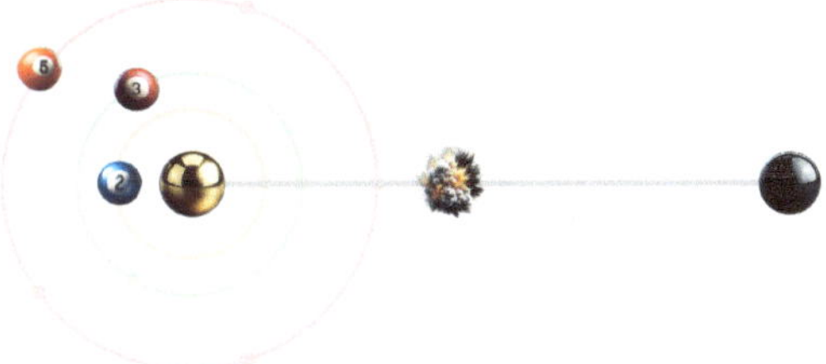

Figure 1.15: Generation of the Fourth Primeon at $t = 7$

pr_4 orbits with circumference 7 units.

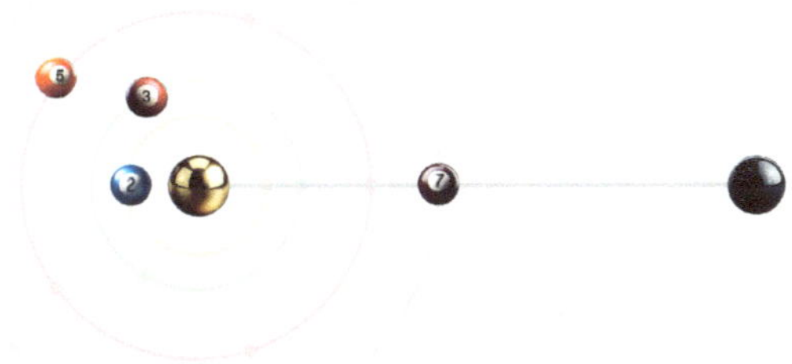

Figure 1.16: Atomic Model of Number 7 (A_7)

1.5 Properties of Atoms

Here is a brief overview of several important properties of numerical atoms, explained with an example.

Consider A_{10}:

Figure 1.17: Atomic Model of Number 10 (A_{10})

A_{10} has four Primeons: $\{pr_1, pr_2, pr_3, pr_4\}$. Two of them, pr_1 and pr_3, are positioned on the critical line, indicating that 2 and 5 are distinct prime factors of 10.

Now compare this with prime atoms such as A_5 or A_7. They have exactly one Primeon on the critical line, and it appears in the last orbit. By examining the atomic structure, we can see whether the atom corresponds to a prime or a composite.

To analyze the structure more precisely, we define the *Atomic Configuration*.

1.5.1 Atomic Configuration

The configuration of an atom is a mathematical representation that uniquely describes the structure of the atom.

Fix a natural number n, and let

$$p_1 = 2, \; p_2 = 3, \; p_3 = 5, \; \ldots, \; p_s$$

be all primes less than or equal to n. The atom A_n contains the Primeons $\{pr_1, pr_2, \ldots, pr_s\}$ with corresponding energy levels $\{2, 3, 5, \ldots, p_s\}$.

At each time t, each Primeon pr_j moves along its orbit of circumference p_j with unit speed. We record the position of pr_j by a number between 0 and $p_j - 1$, where 0 means that the Primeon lies on the critical line.

The configuration of A_n at time t is the list of these positions for all Primeons pr_j with $1 \leq j \leq s$.

Let us examine A_{10} at time $t = 10$. Its Primeons have energy levels $2, 3, 5, 7$, and their positions are:

Primeons	pr_1	pr_2	pr_3	pr_4
Energy Level	2	3	5	7
Positions	0	1	0	3

Table 1.1: **Atomic Configuration of A_{10} at $t = 10$**

So the configuration of $A_{10}(10)$ is $\{0, 1, 0, 3\}$. This configuration has two zeros, meaning two Primeons lie on the critical line, indicating that 2 and 5 are distinct prime factors of 10.

In general, each zero in the configuration indicates a distinct prime factor of the current time t (within the set of Primeons available in the atom). Therefore, in the prime case, we expect no zeros among the earlier Primeons, and only the final-orbit behavior remains special in this model.

1.5.2 Configuration of Numerical Atoms Over Time t

We write $A_n(t)$ to represent the configuration of A_n at time t.

We already observed A_{10} at $t = 10$. So the configuration of $A_{10}(10)$ is $\{0, 1, 0, 3\}$, matching the structure in Figure:

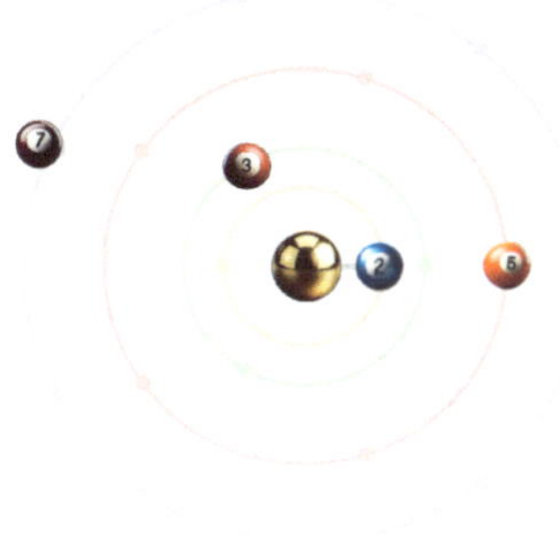

Figure 1.18: Structure of $A_{10}(10)$

Atoms A_7, A_8, A_9, and A_{10} contain the same set of Primeons $\{2, 3, 5, 7\}$. Therefore, at any time t, their configurations match on these shared Primeons.

Now observe the configuration at the next second, $t = 11$:

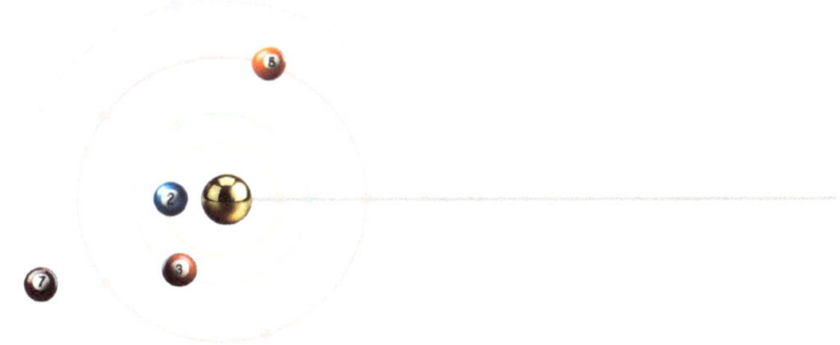

Figure 1.19: Structure at $t = 11$ (for an atom containing Primeons $\{2, 3, 5, 7\}$)

At $t = 11$, the structure has no Primeons on the critical line. The configuration is $\{1, 2, 1, 4\}$, and it contains no zeros. This means 11 has no prime factors among $\{2, 3, 5, 7\}$, and so 11 is the next prime.

Continuing to $t = 12$, the configuration becomes $\{0, 0, 2, 5\}$, indicating that 12 has two distinct prime factors: $\{2, 3\}$. At $t = 13$, the configuration is $\{1, 1, 3, 6\}$, indicating that 13 is prime.

Each atom can perfectly distinguish primes and composites up to a certain range. In the next section, we determine this range.

Let us start with a question:

Question. What is the first composite number that does *not* have any prime factors in $\{2, 3, 5, 7\}$? Please think and continue when you find an answer. Don't worry if it is wrong—just choose a number.

Solution. If we are not allowed to use prime factors in $\{2, 3, 5, 7\}$, then the smallest prime factor we can use is 11. A composite must have at least two prime factors (counting multiplicity), so the smallest such composite is

$$11^2 = 121.$$

In general, the first composite that has no prime factors less than or equal to p_s should be $(p_{s+1})^2$. Therefore, $A_{p_s}(t)$ can perfectly determine whether every number $t < (p_{s+1})^2$ is prime or composite.

We can expand the view range further. Even beyond this boundary, $A_{p_s}(t)$ still checks whether t has any factors in $\{2, 3, \ldots, p_s\}$. It filters out those composites and leaves the remaining integers as more likely to be primes, which we define as *prime candidates*.

Challenge:

1. What are the second and third composite numbers that do not have any prime factors in $\{2, 3, 5\}$?

2. What are the second and third composite numbers that do not have any prime factors in $\{2, 3, 5, 7\}$?

1.5.3 All Atoms Are Synchronized

An important property of atoms is that they are synchronized. We can transfer knowledge from smaller atoms to larger ones and vice versa. Whenever two atoms A_n and A_m both contain a Primeon pr_j, that Primeon has the same position in both configurations at any time t. This synchronization property becomes valuable when we design prime-generating and prime-filtering procedures.

1.5.4 Atomic Periodicity

One of the most important properties of atomic configurations is that all numerical atoms are periodic. In mathematical terms, the configuration of $A_n(t)$ repeats after a fixed time called the period T_n:

$$A_n(t + T_n) = A_n(t).$$

For example, the period of A_3 is $T_3 = 6$:

Figure 1.20: (A_3) Has a Period of $T_3 = 6$

This means that for every t,

$$A_3(t + 6) = A_3(t).$$

Periodicity is one of the most important concepts in mathematics. In the next chapter, we will focus on understanding periods through remainders.

1.5.5 Primeon Wave

One of the most important properties in the Atomic Model of Numbers is that Primeons exhibit a wave-particle duality inside this framework. They can be treated as particles (as in the pictures) or as waves (as in periodic motion and phase). We will analyze configurations from both perspectives and develop additional tools for understanding the structure behind primes.

1.6 Future Chapters

The Atomic Model of Numbers narrates the generation of primes through a simple mechanism. The key lies in a deep understanding of **remainders**. Mastering the calculation and properties of remainders is essential to unraveling the mechanics of this model. Our next goal is to uncover remainders, learn their patterns, and seek efficient ways to compute them.

Chapter 2

Remainder Calculation Sequences

"Prime numbers are more numerous than any assigned multitude of prime numbers." — *Euclid*

Figure 2.1: Euclid (300 BCE)

2.1 Preface

2.1.1 Introduction

Prime numbers are among the most beautiful concepts in mathematics. A prime number has exactly two positive divisors: 1 and itself. In other words, dividing a prime by any integer smaller than the prime and greater than 1 will always leave a positive remainder.

The concept of primes is closely related to remainders. In this chapter, we will explore how to calculate remainders efficiently. We will focus on the concept of *period*, because recognizing periodicity is key to identifying patterns. We will introduce **Remainder Calculation Sequences** (RCS), which are periodic sequences used to compute remainders. Before diving into these topics, we first equip ourselves with some fundamental tools in number theory.

We dedicate this chapter to Euclid and Āryabhaṭa—pioneers whose groundbreaking work in mathematics, geometry, and number theory laid much of the foundation for these explorations.

Euclid

Euclid is widely revered as one of history's greatest mathematicians, known as the "Father of Geometry." Living around 300 BCE in Alexandria, he wrote the influential treatise *The Elements*, which systematically developed geometry from basic definitions and postulates. This work introduced rigorous, proof-based reasoning that set the standard for mathematics for more than two millennia. Beyond geometry, Euclid explored number theory, including prime numbers, perfect numbers, and the Euclidean algorithm—concepts that remain central to mathematics today. His famed statement, "There is no royal road to geometry," reflects his commitment to logical rigor. Though little is known of his personal life, Euclid's lasting influence continues to shape mathematics, philosophy, science, and education, exemplifying humanity's enduring quest to understand the universe through reason and inquiry.

Āryabhaṭa

Āryabhaṭa (c. 476–550 CE) was one of the earliest and most influential mathematician–astronomers of the classical Indian tradition, best known for his compact treatise *Āryabhaṭīya* (499 CE), which shaped mathematical astronomy for centuries. In it he presented systematic methods for arithmetic, algebraic manipulation, and trigonometry—introducing a sine-based approach (using half-chords) and practical procedures for constructing tables and solving computational problems. He gave sophisticated algorithms for extracting square and cube roots, worked with place-value notation in a highly operational way, and produced remarkably accurate astronomical parameters for his time, including a close estimate of π and an explanation of eclipses consistent with a shadow model rather than mythic causes. Āryabhaṭa's blend of terse verse, algorithmic thinking, and quantitative astronomy became a foundation that later scholars expanded into major schools of Indian mathematics and celestial calculation.

Challenge

Challenge 1: Periodic Digits Function

Let's define a function $P(\alpha, \ell)$ that repeats the digits of α until the total length is ℓ. For example:

$$P(123, 5) = 12312, \quad P(20, 7) = 2020202, \quad P(12345, 3) = 123.$$

Find a mathematical function for $P(\alpha, \ell)$.

Challenge 2: Calculate Remainders to 7 for Big Numbers

Based on the defined periodic digits function, calculate the remainder of this huge number divided by 7:

$$P(123456789, 10^{123456789}) \bmod 7.$$

Challenge 3: $c^2 = \dfrac{a^2\, b^2}{a^2 + b^2}$

2.2 Fundamentals of Number Theory

2.2.1 Fundamentals of Numbers

In this section, we become familiar with different types of numbers. Let's begin the story of numbers with zero and one, viewed as symbols of non-existence and existence. They give meaning to each other; without the other, each is meaningless.

The simplest operation is **addition**. By repeatedly adding one, we generate an infinite sequence of *natural numbers*:

$$0 + 1 = 1, \quad 1 + 1 = 2, \quad 2 + 1 = 3, \quad 3 + 1 = 4, \quad \ldots$$

$$\mathbb{N} = \{1, 2, 3, 4, \ldots\}.$$

Functions are expressions formed by combining variables, constants, operations, and possibly other functions.

The simplest function is the addition function, or sum function. The inverse of the addition function is **subtraction**: if a function sends x to y, its inverse takes y and returns x.

When we subtract a larger number from a smaller one, we encounter *negative* numbers. To include these, we expand our number system to the *integers*:

$$\mathbb{Z} = \{\ldots, -2, -1, 0, 1, 2, \ldots\}.$$

Next comes **multiplication**, which we view as repeated addition:

$$a + a + a + a + a = 5 \times a.$$

The inverse of multiplication is **division**:

$$\frac{5 \times a}{5} = a.$$

In a fraction, the top is the numerator and the bottom is the denominator. Not every numerator divides evenly by its denominator. For example, if you divide 29 by 5:

$$\frac{29}{5} = 5.8,$$

which is not an integer. This leads us to *rational numbers* ($\mathbb{Q}$), where each number can be written as $\frac{a}{b}$ (with $a, b \in \mathbb{Z}$, $b \neq 0$).

The rational number $\frac{29}{5}$ can be split as

$$\frac{29}{5} = \frac{25}{5} + \frac{4}{5} = 5 + \frac{4}{5}.$$

Hence, dividing 29 by 5 yields a quotient of 5 and a remainder of 4.

All rational numbers can be written as decimals, and these decimals are ultimately periodic. For instance, $\frac{1}{7}$ has a period of 6:

$$\frac{1}{7} = 0.\overline{142857},$$

while $\frac{1}{8}$ has a repeating period of 1 on the digit 0:

$$\frac{1}{8} = 0.125\overline{0}.$$

Some numbers like π, $\sqrt{2}$, or $\log(2)$ cannot be written as fractions or as decimals with fixed periods. Although they differ in nature, we call them *irrational* numbers.

Together, rational and irrational numbers form the set of *real numbers* ($\mathbb{R}$).

By repeated multiplication, we define the **exponentiation** function:

$$f(x) = b^x = y,$$

where b is the base and x is the exponent. For example,

$$f(x) = 2^x.$$

The inverse of exponentiation is the **logarithm** function, which takes the output and the base, returning the exponent:

$$\log_b(y) = x.$$

Logarithms can be understood as repeated divisions by the base until reaching 1.

Example: Calculate $\lfloor \log_3(100) \rfloor$.

Divide 100 by 3 repeatedly:

$$100/3 \approx 33.3333,$$

$$(100/3)/3 \approx 11.1111,$$

$$((100/3)/3)/3 \approx 3.7037,$$

$$(((100/3)/3)/3)/3 \approx 1.2346,$$

$$((((100/3)/3)/3)/3)/3 \approx 0.4115.$$

After 4 divisions, we are still above 1; after 5 divisions, we drop below 1, so $\log_3(100)$ is between 4 and 5. Thus $\lfloor \log_3(100) \rfloor = 4$.

We can also multiply from 1 up to 100 to find the same result:

$$1 \times 3 = 3, \quad 3 \times 3 = 9, \quad 9 \times 3 = 27, \quad 27 \times 3 = 81, \quad 81 \times 3 = 243.$$

Since $3^4 \leq 100 < 3^5$, it follows again that $\lfloor \log_3(100) \rfloor = 4$.

Next, consider the power function:

$$f(x) = x^p = y,$$

where x is the variable and p is the constant exponent. For example,

$$f(x) = x^2.$$

Its inverse is the root function:

$$\sqrt[p]{y} = x.$$

The solution of $x^2 = -1$ is $x = \sqrt{-1}$, which does not exist in the real number system. Defining $\sqrt{-1}$ as the imaginary unit extends our system to *complex numbers*, where

$$z = a + i\,b, \quad (a, b \in \mathbb{R}).$$

using complex system we can greatly formulate the Atomic Model of Numbers. In this seasson we promote simplicity and conceptual understanding and we only learn Fundamental of Number Theory and Algebra. If this book continues In future seassons we will cover Infinite Series, Calculus, Complex System and we will exapnd

This was a short overview of the main number systems we use.

2.2.2   Divisibility and Divisors

Definition (Divisibility).   An integer a is **divisible** by another integer b (where $b \neq 0$) if there is no remainder when a is divided by b. Symbolically:

$$b \mid a \iff a = b \times k \quad \text{(for some integer } k\text{)}.$$

Example.   21 is divisible by 7 because $21 = 7 \times 3$.

Set of Divisors.   All integers d such that $d \mid n$ form the set of divisors of n:

$$\text{Divisors of } 21 = \{1, 3, 7, 21\}.$$

2.2.3   Common Divisors and GCD

Common Divisors.   If two integers x and y share a divisor d, then d is a **common divisor** of x and y. For example:

$$12 : \{1, 2, 3, 4, 6, 12\}, \quad 21 : \{1, 3, 7, 21\} \implies \text{Common divisors} = \{1, 3\}.$$

Coprime. $\gcd(12, 35) = 1$ means 12 and 35 share no prime factors. They are called **coprime**.

Greatest Common Divisor (GCD). $\gcd(a, b)$ is the largest integer dividing both a and b. For instance, $\gcd(12, 18) = 6$.

Euclidean Algorithm.   To find the GCD using the Euclidean Algorithm, note that if $a = b\,k + r$ and d divides both a and b, then d also divides r. Here, r is the remainder of a divided by b.
A fast way to find gcd:

$$\gcd(48, 18) = \gcd(18,\ 48 \bmod 18) = \gcd(18, 12) = \gcd(12, 6) = \gcd(6, 0) = 6.$$

2.2.4 Least Common Multiple (LCM)

Definition (LCM). $\mathrm{lcm}(a, b)$ is the smallest positive integer divisible by both a and b. For example, $\mathrm{lcm}(6, 15) = 30$.

2.2.5 Period

A *period* is an important concept in mathematics: many patterns repeat after a fixed number of steps. The LCM of a set of numbers gives the combined period of those numbers.

Example. To calculate the LCM of $\{3, 4\}$:

$$
\begin{array}{ccccccccccccc}
1 & 2 & 3 & 4 & 1 & 2 & 3 & 4 & 1 & 2 & 3 & 4 & 1\dots \\
1 & 2 & 3 & 1 & 2 & 3 & 1 & 2 & 3 & 1 & 2 & 3 & 1\dots
\end{array}
$$

The period here is 12. After 12 steps from $\begin{pmatrix} 1 \\ 1 \end{pmatrix}$, we return to $\begin{pmatrix} 1 \\ 1 \end{pmatrix}$.

Similarly, for $\{2, 5\}$:

$$
\begin{array}{ccccccccccc}
1 & 2 & 1 & 2 & 1 & 2 & 1 & 2 & 1 & 2 & 1 \\
1 & 2 & 3 & 4 & 5 & 1 & 2 & 3 & 4 & 5 & 1
\end{array}
$$

The combined period is 10.

And for $\{4, 6, 12\}$:

$$
\begin{array}{ccccccccccccc}
1 & 2 & 3 & 4 & 1 & 2 & 3 & 4 & 1 & 2 & 3 & 4 & 1 \\
1 & 2 & 3 & 4 & 5 & 6 & 1 & 2 & 3 & 4 & 5 & 6 & 1 \\
1 & 2 & 3 & 4 & 5 & 6 & 7 & 8 & 9 & 10 & 11 & 12 & 1
\end{array}
$$

The combined period is 12.

GCD and LCM Key Relationship. For two numbers a and b:

$$
\gcd(a, b) \times \mathrm{lcm}(a, b) = a \times b.
$$

To calculate GCD and LCM by factorization, factorize all numbers into primes:

- **GCD**: the product of all shared prime factors with their minimum exponents.

- **LCM**: the product of all unique prime factors with their maximum exponents.

For example, for $\{4, 6, 12\}$:

$$
4 = 2^2,\ 6 = 2 \times 3,\ 12 = 2^2 \times 3.
$$

$$
\gcd(4, 6, 12) = 2^{\min(2,1,2)} = 2^1 = 2.
$$

$$
\mathrm{lcm}(4, 6, 12) = 2^{\max(2,1,2)} \times 3^{\max(0,1,1)} = 2^2 \times 3 = 12.
$$

2.2.6   Remainders and Modulo Arithmetic

Remainder.   When dividing a by a positive integer b:

$$a = bq + r, \quad 0 \le r < b.$$

The expression mod represents the remainder:

$$a \bmod b = r.$$

For example, $17 = 5 \times 3 + 2$ implies $17 \bmod 5 = 2$.

Here, r can be any of $\{0, 1, 2, \ldots, b-1\}$.

Sometimes we use *symmetric modulo*, where the remainder is centered around zero:

$$a \bmod {}_s b = r, \quad r \in \left\{ -\left\lfloor \frac{b}{2} \right\rfloor, \ldots, 0, \ldots, \left\lfloor \frac{b}{2} \right\rfloor \right\}.$$

Congruence.   We say $a \equiv b \pmod{m}$ if a and b leave the same remainder when divided by m. For example, $17 \equiv 2 \pmod 5$.

2.3   Remainder Calculation Sequences (RCS)

2.3.1   Calculating Remainders to 7

Calculate the remainder:

$$31415926535 \bmod 7.$$

Although 7 is a small number, directly computing such a large remainder can be cumbersome without a method. A useful trick is to remember three numbers in order: 1, 3, and 2. There is a periodic sequence that starts with these three numbers and then continues with their negatives. This periodic sequence is

$$[1, 3, 2, -1, -3, -2],$$

which we call the *Remainder Calculation Sequence* (RCS) of 7.

To calculate a remainder, write the sequence under each digit of the number from right to left, multiply each digit by its corresponding sequence value, and sum the products:

$$
\begin{array}{rrrrrrrrrrr}
3 & 1 & 4 & 1 & 5 & 9 & 2 & 6 & 5 & 3 & 5 \\
-3 & -1 & 2 & 3 & 1 & -2 & -3 & -1 & 2 & 3 & 1 \\
-9 & -1 & +8 & +3 & +5 & -18 & -6 & -6 & +10 & +9 & +5
\end{array}
$$

Adding these results:

$$(-9 - 1 + 8 + 3 + 5 - 18 - 6 - 6 + 10 + 9 + 5) \bmod 7 = 0 \bmod 7 = 0.$$

Therefore, the remainder is 0.

Alternatively, you can arrange a table, putting $[1, 3, 2]$ at the top (from right to left) and grouping the large number in 3-digit blocks from the right side, alternating signs for each row. You then sum each column and multiply by the header. In every approach, you end up with the same result, 0.

RCS.   Remainder Calculation Sequences are periodic sequences used for computing remainders. They usually have two parts: initial steps and a periodic core.

For small numbers, it is handy to memorize these sequences.

Number	Initial Steps	Periodic Sequence
1	[]	[0]
2	[1]	[0]
3	[]	[1]
4	[1, 2]	[0]
5	[1]	[0]
6	[1]	[-2]
7	[]	[1, 3, 2, -1, -3, -2]
8	[1, 2, -4]	[0]
9	[]	[1]
10	[1]	[0]
11	[]	[1, -1]
12	[1, -2]	[4]
13	[]	[1, -3, -4, -1, 3, 4]

Examples.

$$123456789 \bmod 11.$$

Here, $\mathrm{RCS}_{11} = [\,[]\,, \, [1, -1]\,]$. Align digits and sequence:

$$
\begin{array}{ccccccccc}
1 & 2 & 3 & 4 & 5 & 6 & 7 & 8 & 9 \\
1 & -1 & 1 & -1 & 1 & -1 & 1 & -1 & 1
\end{array}
$$

$$(1 - 2 + 3 - 4 + 5 - 6 + 7 - 8 + 9) \bmod 11 = 5.$$

Another example:

$$123456789 \bmod 8.$$

From RCS_8, the initial steps are $[1, 2, -4]$ and then a periodic sequence of 0. Evaluating:

$$(7 \times (-4) + 8 \times 2 + 9 \times 1) \bmod 8 = (-28 + 16 + 9) \bmod 8 = -3 \bmod 8 = 5.$$

2.3.2   Generating RCS

Remembering RCS for small n is useful. For larger n, since RCS is periodic, it is often beneficial to calculate the RCS first and then use it to compute remainders, rather than performing direct long division.

To generate RCS_n:

$$\mathrm{RCS}_n = \left[10^0 \bmod n, \ 10^1 \bmod n, \ 10^2 \bmod n, \dots \right].$$

Compute terms until you return to a previously seen value, at which point you've identified both the initial steps and the periodic part. Using symmetric modulo (centered around zero) often simplifies arithmetic:

$$\mathrm{RCS}_n = \left[10^0 \bmod {}_s n, \ 10^1 \bmod {}_s n, \ 10^2 \bmod {}_s n, \dots \right].$$

Example: Calculate $31415926535 \bmod 21$.
First find $10^x \bmod 21$:

$$10^0 \equiv 1, \ 10^1 \equiv 10, \ 10^2 \equiv 16, \ 10^3 \equiv 13, \ 10^4 \equiv 4, \ 10^5 \equiv 19, \ 10^6 \equiv 1 \quad (\bmod 21).$$

We returned to 1, so the period is $[1, 10, 16, 13, 4, 19]$. Using symmetric modulo, we can write RCS_{21} as $[1, 10, -5, -8, 4, -2]$.

Arrange the digits of 31415926535 with these RCS values:

3	1	4	1	5	9	2	6	5	3	5
4	-8	-5	10	1	-2	4	-8	-5	10	1
12	-8	-20	$+10$	$+5$	-18	$+8$	-48	-25	$+30$	$+5$

Summing these results and taking mod 21:

$$-49 \bmod 21 = 14.$$

So, $31415926535 \bmod 21 = 14$.

2.4   Solving Challenges

Periodic Digits

Goal.   The purpose of this challenge is to engage you with numbers, stimulate your observation skills, and deepen your understanding of periods and remainder calculations.

We want a function $P(\alpha, \ell)$ that repeats the digits of α until reaching a total length ℓ.

Decimal expansions can be periodic. We observe this pattern by looking at fractions like:

$$\frac{1}{9} = 0.\overline{1}.$$

Also:

$$\frac{1}{99} = 0.\overline{01}, \quad \frac{1}{999} = 0.\overline{001}.$$

We see that if we want a period equal to α, we can use

$$\frac{\alpha}{\underbrace{99\ldots9}_{\text{digits in }\alpha}}.$$

For example, $0.\overline{123} = \frac{123}{999}$, $0.\overline{1234} = \frac{1234}{9999}$.

Thus, if α is a natural number,

$$\frac{\alpha}{10^{\lfloor \log_{10}(\alpha) \rfloor + 1} - 1} = 0.\overline{\alpha}.$$

Multiplying by 10^{ℓ} and applying the floor function yields:

$$P(\alpha, \ell) = \left\lfloor 10^{\ell} \cdot \frac{\alpha}{10^{\lfloor \log_{10}(\alpha) \rfloor + 1} - 1} \right\rfloor.$$

Calculate Remainders to 7 for Big Numbers

Problem 2. Calculate
$$P(123456789, 10^{123456789}) \bmod 7.$$

The number $10^{123456789}$ is enormous, so direct calculation is impossible. Instead, we use the RCS of 7, which has a period of 6. The number 123456789 has 9 digits, so their combined period is $\mathrm{lcm}(6, 9) = 18$.

We then calculate
$$10^{123456789} \bmod 18.$$

All numbers of the form $10^n - 1$ are divisible by 9. Considering even and odd exponents, we find patterns modulo 18 that simplify the analysis. Essentially, we discover that
$$10^n \bmod 18 = 10.$$

Thus, $P(123456789, 10^{123456789})$ repeats in blocks of 18 digits, followed by 10 extra digits at the end. We line these digits up with their corresponding RCS values (from right to left).

Upon careful summation, each 18-digit block sums to 0 in mod 7. The final 10 digits sum to 25, giving
$$25 \bmod 7 = 4.$$

Therefore,
$$P(123456789, 10^{123456789}) \bmod 7 = 4.$$

Note that elements of RCS_n are not unique. We can choose from positive or negative options. For our example, if r is in RCS_n, then $r + 7k$ for any integer k is also valid since $r \equiv r + 7k \pmod 7$.

As an optimization in sequential calculations, you can keep partial sums near zero by choosing suitable representatives from the RCS.

Challenge 3: Solutions for $c^2 = \dfrac{a^2\, b^2}{a^2 + b^2}$

Problem 3. $c^2 = \dfrac{a^2\, b^2}{a^2 + b^2}$

Find a solution for (a, b, c, n) in the equation
$$c^2 = \frac{a^2\, b^2}{a^2 + b^2}.$$

Assume $n = 2$. Invert both sides:
$$\frac{1}{c^2} = \frac{1}{a^2} + \frac{1}{b^2}.$$

This is reminiscent of the Pythagorean theorem in reverse:
$$z^2 = x^2 + y^2.$$

Take the classic right triangle with $x = 3$, $y = 4$, and $z = 5$. Divide all sides by $(xyz)^2$,
$$\frac{1}{(xy)^2} = \frac{1}{(yz)^2} + \frac{1}{(xz)^2}.$$

Compare to match:
$$a = xz, \quad b = yz, \quad c = xy.$$

Hence $a = 15$, $b = 20$, and $c = 12$, with $n = 2$.

2.5 Future Challenges

We defined
$$\mathrm{RCS}_n = \left[\, 10^0 \bmod n,\ 10^1 \bmod n,\ 10^2 \bmod n, \ldots \right],$$

which has two parts: initial steps and a periodic sequence. From the RCS table, we wonder about the length of the initial steps and the period. Since $f(x) = 10^x \bmod n$ is an exponential function, we need to understand the properties of exponential remainders. In the next chapter, we will learn algebra and then apply our knowledge of number theory to find efficient ways to calculate $a^b \bmod c$.

2.6 Conclusion

Understanding prime numbers requires a solid grasp of remainders. In this chapter, we introduced a challenge-solving strategy to calculate remainders, placing special emphasis on periods. We created the periodic digit function to build large numbers with repetitive patterns. We also defined Remainder Calculation Sequences (RCS) and showcased their power in solving remainder problems. However, new questions arise: How do we efficiently calculate the initial step size and period, and how do we compute exponential remainders? These questions set the stage for the next chapters.

Chapter 3

Fundamentals of Algebra

"Truth is ever to be found in simplicity, and not in the multiplicity and confusion of
things."
— Isaac Newton

Figure 3.1: Al-Khwarizmi (780–850)

3.1 Preface

3.1.1 Introduction

The discovery of mathematics began with measuring physical objects. However, mathematics itself is not bound to any object—objects are merely variables. One of the great revolutionary stages in mathematics was the separation of mathematics from physics.

In fact, mathematics is an abstract science that governs the rules and behavior of objects. Objects have limitations, but when mathematics is separated from them, it finds freedom: the freedom to travel between worlds and spaces, and to extend toward infinity.

One of the greatest steps in this journey came through the work of Al-Khwarizmi, the "Father of Algebra," who revolutionized the field and ushered in an era of pure mathematical thought. After Khwarazmi AlK araji

We dedicate this chapter to Al-Khwarizmi and Al-Karaji, for laying essential foundations in algebraic thinking and computational methods that power the discoveries ahead.

Al-Khwarizmi (Khwarazmi): The Father of Algebra

Muhammad ibn Musa al-Khwarizmi (c. 780 – c. 850) was a Persian mathematician and scholar whose work shaped the foundations of modern mathematics. He developed systematic methods for solving equations, which earned him the title "Father of Algebra." The term "algorithm" is derived from his name, reflecting his lasting influence on mathematics and computation. His contributions also advanced astronomy, geography, and cartography, making him one of history's most influential scientific minds.

Al-Karaji: Extending Algebra Beyond Equations

Abu Bakr Muhammad ibn al-Husayn al-Karaji (fl. c. 1000) carried algebra forward from "solving equations" into "building algebraic structure." He treated polynomials as objects that can be manipulated by rules—expanded, factored, compared, and transformed—so that algebra becomes a language for patterns, not only a tool for answers.

A central theme in Al-Karaji's work is *generalization*: formulas for families of expressions, identities that scale with degree, and arguments that climb step-by-step from one case to the next. This mindset naturally connects to what we now call *mathematical induction*, which appears in early forms in his treatment of power sums and related identities.

Al-Karaji's influence also extends beyond pure algebra. He wrote on applied mathematics and engineering—most famously on hidden waters and groundwater extraction—showing the same algebraic spirit: define a structure, identify invariants, and work by systematic reasoning rather than isolated examples.

Figure 3.2: Al-Karaji (fl. c. 1000)

3.2 Challenge

Imagine a circle of radius r centered at the origin. The goal is to count how many integer points (x, y) lie within or on the boundary of this circle, where each point satisfies the inequality

$$x^2 + y^2 \leq r^2.$$

The integer points (x, y) that meet this condition are called **lattice points**. These points are located in all four quadrants of the circle.

Your challenge is twofold:

1. **Exact formula:** Derive a formula that computes the total number of lattice points, denoted by $N(r)$, for a given radius r.

2. **Approximation formula:** Derive an approximation formula for $N(r)$.

Example: Consider $r = 5$. We want to find all lattice points (x, y) such that

$$x^2 + y^2 \leq 25.$$

The lattice points that satisfy this for $r = 5$ include points such as $(0, 0)$, $(3, 4)$, $(-3, 4)$, $(5, 0)$, and many others. There are exactly 81 such points.

Visual representation for $r = 5$:

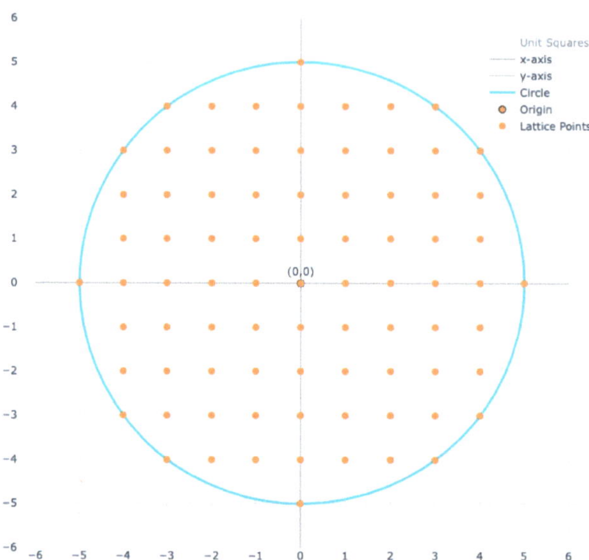

Figure 3.3: Gauss circle lattice points for a circle with radius $r = 5$

3.3 Algebra

3.3.1 Definitions and Basic Concepts

Variables

A **variable** is a symbol, usually a letter, that represents one or more numbers. Variables allow us to write general expressions and equations that hold true for many values. For example, in the equation $x + 2 = 5$, the symbol x is a variable representing the unknown number we wish to find.

Constants

A **constant** is a fixed value that does not change. In equations and expressions, constants are known numerical values. For example, in $x + 2 = 5$, the numbers 2 and 5 are constants.

Expressions

An **algebraic expression** is a combination of variables, constants, and operators (such as $+, -, \times, \div$). Expressions represent quantities and can be simplified or evaluated given specific variable values. For example, $3x + 2$ is an algebraic expression.

Equations

An **equation** is a mathematical statement that asserts the equality of two expressions. It consists of two expressions separated by an equals sign ($=$). Solving an equation involves finding the value(s) of the variable(s) that make the statement true. For example, $2x + 3 = 7$ is an equation.

Example: Solve $2x + 3 = 7$.

Solution:

$$2x + 3 = 7,$$
$$2x = 7 - 3 = 4,$$
$$x = \frac{4}{2} = 2.$$

Answer: $x = 2$.

Inequalities

An **inequality** is a mathematical statement that compares two expressions, indicating that one is greater than, less than, greater than or equal to, or less than or equal to the other. Symbols used include $>, <, \geq, \leq$. For example, $x + 3 > 5$ is an inequality.

Absolute Value

The **absolute value** of a number is its distance from zero on the number line, regardless of direction. It is always non-negative and is denoted by $|x|$. For example, $|5| = 5$ and $|-3| = 3$.

Properties of Absolute Value

- **Non-negativity:** $|x| > 0$.

- **Identity:** $|0| = 0$.

- **Symmetry:** $|-x| = |x|$.

- **Multiplicative property:** $|xy| = |x|\,|y|$.

- **Triangle inequality:** $|x + y| \leq |x| + |y|$.

Example: Solve $|x - 3| = 5$.

Solution:

An absolute value equation $|A| = B$ implies $A = B$ or $A = -B$.

$$x - 3 = 5 \quad \text{or} \quad x - 3 = -5.$$

Solving both equations:

$$x - 3 = 5 \quad \Rightarrow \quad x = 8,$$
$$x - 3 = -5 \quad \Rightarrow \quad x = -2.$$

Answer: $x = 8$ or $x = -2$.

3.3.2 Operations and Properties

Addition and Subtraction

- **Commutative property of addition**: $a + b = b + a$.

- **Associative property of addition**: $(a + b) + c = a + (b + c)$.

- **Additive identity**: $a + 0 = a$.

- **Additive inverse**: $a + (-a) = 0$.

Multiplication and Division

- **Commutative property of multiplication**: $a \times b = b \times a$.

- **Associative property of multiplication**: $(a \times b) \times c = a \times (b \times c)$.

- **Multiplicative identity**: $a \times 1 = a$.

- **Multiplicative inverse**: $a \times \frac{1}{a} = 1$ (for $a \neq 0$).

- **Distributive property**: $a(b + c) = ab + ac$.

3.3.3 Equations

Linear Equations

A **linear equation** is an equation of the first degree, meaning it has no exponents greater than one. The general form of a linear equation in one variable is

$$ax + b = 0,$$

where a and b are constants, and $a \neq 0$.

    Example: Solve $2x + 3 = 7$.

    Solution:

$$2x + 3 = 7,$$
$$2x = 4,$$
$$x = 2.$$

Answer: $x = 2$.

Quadratic Equations

A **quadratic equation** is a second-degree polynomial equation in a single variable x with the general form

$$ax^2 + bx + c = 0,$$

where a, b, c are constants and $a \neq 0$.

    Example: Solve $x^2 - 5x + 6 = 0$.

    Solution:

$$x^2 - 5x + 6 = 0,$$
$$(x - 2)(x - 3) = 0,$$
$$x - 2 = 0 \quad \text{or} \quad x - 3 = 0,$$
$$x = 2 \quad \text{or} \quad x = 3.$$

Answer: $x = 2$ or $x = 3$.

Systems of Equations

A **system of equations** is a set of two or more equations with the same variables. The solution to the system is the set of variable values that satisfy all equations simultaneously.

Example: Solve the system

$$\begin{cases} x + y = 3, \\ 2x - y = 0. \end{cases}$$

Solution:

Add the two equations to eliminate y:

$$(x + y) + (2x - y) = 3 + 0,$$
$$3x = 3,$$
$$x = 1.$$

Substitute $x = 1$ into the first equation:

$$1 + y = 3 \implies y = 2.$$

Answer: $x = 1, y = 2$.

3.3.4 Polynomials

Definition

A **polynomial** is an expression consisting of variables and coefficients, involving only the operations of addition, subtraction, multiplication, and non-negative integer exponents of variables. The general form of a polynomial in one variable x is

$$P(x) = a_n x^n + a_{n-1} x^{n-1} + \cdots + a_1 x + a_0,$$

where n is a non-negative integer, and $a_n, a_{n-1}, \ldots, a_0$ are constants with $a_n \neq 0$.

Degree of a Polynomial

The **degree** of a polynomial is the highest power of the variable in the polynomial with a non-zero coefficient. For example, the degree of $4x^3 + 3x^2 - 2x + 7$ is 3.

Factoring Polynomials

Factoring a polynomial is the process of expressing the polynomial as a product of its factors. Factoring simplifies polynomials and is useful for solving polynomial equations and finding roots.

Example: Factor $x^2 - 5x + 6$.

Solution:

Find two numbers m and n such that

$$m \times n = 6 \quad \text{and} \quad m + n = -5.$$

The numbers are $m = -2$ and $n = -3$. Therefore,

$$x^2 - 5x + 6 = (x - 2)(x - 3).$$

Common factoring methods

1. **Greatest common factor (GCF)**

   Factor $3x^2 + 6x$.

   Solution:

   The GCF of $3x^2$ and $6x$ is $3x$. Factor out $3x$:
   $$3x^2 + 6x = 3x(x + 2).$$

2. **Factoring by grouping**

   Factor $x^3 - x^2 + 3x - 3$.

   Solution:

   Group terms:
   $$(x^3 - x^2) + (3x - 3).$$
   Factor out common factors in each group:
   $$x^2(x - 1) + 3(x - 1).$$
   Factor out $(x - 1)$:
   $$(x - 1)(x^2 + 3).$$

3. **Difference of squares**

   Factor $x^2 - 9$.

   Solution:

   Recognize that $9 = 3^2$:
   $$x^2 - 3^2 = (x - 3)(x + 3).$$

4. **Sum and difference of cubes**

   Factor $x^3 - 8$.

   Solution:

   Recognize that $8 = 2^3$:
   $$x^3 - 2^3 = (x - 2)(x^2 + 2x + 4).$$

5. **Trinomials**

   Factor $x^2 + 7x + 10$.

   Solution:

   Find two numbers m and n such that
   $$m \times n = 10 \quad \text{and} \quad m + n = 7.$$
   The numbers are $m = 2$ and $n = 5$. Therefore,
   $$x^2 + 7x + 10 = (x + 2)(x + 5).$$

6. **Perfect square trinomials**

   Factor $x^2 + 6x + 9$.

   Solution:

   Recognize that $9 = 3^2$ and $6x = 2 \cdot 3x$. Therefore,
   $$x^2 + 6x + 9 = (x + 3)^2.$$

Factoring Higher-Order Equations

All polynomial equations of degree n yield n roots (counted with multiplicity), but not all roots are real numbers. Even finding real roots can be difficult and generally requires structured methods, especially for degree 2 and degree 3 equations.

The simplest form of polynomial equations are *Diophantine equations*, where all roots are integers. We can use the properties of prime numbers to solve these equations step by step.

Example 1. Assume all roots are integers. Find the roots of the cubic equation

$$x^3 + 11x^2 + 31x + 21 = 0.$$

Solution:

When dealing with higher-order equations, it is useful to simplify the equation whenever we recognize a root.

The first step is to test $x = 0$, then $x = -1$ and $x = 1$ in the equation. If you find a root, factor it out and repeat the process on the remaining factor.

Observe that $x = -1$ is a root because

$$(-1)^3 + 11(-1)^2 + 31(-1) + 21 = 0.$$

Now, factor out $(x - (-1)) = (x + 1)$ from the equation by grouping:

$$\begin{aligned}
x^3 + 11x^2 + 31x + 21 &= (x^3 + x^2) + (10x^2 + 10x) + (21x + 21) \\
&= x^2(x + 1) + 10x(x + 1) + 21(x + 1) \\
&= (x + 1)(x^2 + 10x + 21).
\end{aligned}$$

Now consider $x^2 + 10x + 21$. Recall the identity for reconstructing a quadratic from its roots:

$$(x - r_1)(x - r_2) = x^2 - (r_1 + r_2)x + r_1 r_2.$$

We set up the system:

$$r_1 + r_2 = -10,$$
$$r_1 r_2 = 21.$$

The factorization $21 = 3 \times 7$ and the sum -10 suggest both roots are negative. Thus $r_1 = -3$ and $r_2 = -7$.

Therefore,

$$x^3 + 11x^2 + 31x + 21 = (x + 1)(x + 3)(x + 7).$$

Example 2. Consider

$$x^3 + 13x^2 + 46x + 48.$$

The equation $x^3 + 13x^2 + 46x + 48 = 0$ has no roots in $\{-1, 0, 1\}$.

We use the expansion

$$(x - r_1)(x - r_2)(x - r_3) = x^3 - (r_1 + r_2 + r_3)x^2 + (r_1 r_2 + r_1 r_3 + r_2 r_3)x - r_1 r_2 r_3,$$

which leads to the system

$$r_1 + r_2 + r_3 = -13,$$
$$r_1 r_2 + r_1 r_3 + r_2 r_3 = 46,$$
$$r_1 r_2 r_3 = -48.$$

From the third equation, the product of the three roots is negative, so either all roots are negative or exactly one of them is negative. The first equation shows that their sum is negative, which suggests that either all roots are negative or one is significantly more negative than the other two.

Next, assume R_1, R_2, R_3 are the magnitudes of the roots. Factorizing

$$48 = 2^4 \times 3,$$

we must assign the prime factors among the three roots, with each root containing at least one prime factor.

Let us parametrize:

$$R_1 = 3t_1, \quad R_2 = 2t_2, \quad R_3 = 2t_3,$$

so that

$$t_1 t_2 t_3 = 2^2.$$

Possible assignments for (t_1, t_2, t_3) include:

$$t_1 = 4 : \ R_1 = 12, \ R_2 = 2, \ R_3 = 2,$$
$$t_1 = 2 : \ R_1 = 6, \ R_2 = 4, \ R_3 = 2,$$
$$t_1 = 1 : \ (R_1 = 3, \ R_2 = 4, \ R_3 = 4) \text{ or } (R_1 = 3, \ R_2 = 8, \ R_3 = 2).$$

Considering the signs based on the earlier discussion, the possible sets of roots (r_1, r_2, r_3) include

$$(-12, -2, -2), \ (-12, 2, 2), \ (-6, -4, -2), \ (-6, 4, 2),$$
$$(-3, -4, -4), \ (3, 4, -4), \ (-3, -8, -2), \ (3, -8, 2).$$

Test these in the equation $r_1 + r_2 + r_3 = -13$:

$$-3 - 8 - 2 = -13.$$

This satisfies the equation, so $(-3, -8, -2)$ are the roots (up to permutation).

Thus,

$$x^3 + 13x^2 + 46x + 48 = (x + 2)(x + 3)(x + 8).$$

Challenge: Solve the following equations

Polynomials only:

1. $x^3 + 12x^2 + 32x,$

2. $x^3 + 12x^2 + 29x + 18,$

3. $x^3 + 8x^2 + 7x,$

4. $x^3 + 10x^2 + 31x + 30.$

Use these problems to practice identifying roots through inference and subsequently factorizing the given cubic equations.

3.3.5 Common Algebraic Identities

1. **Square of a sum**
$$(x + y)^2 = x^2 + 2xy + y^2.$$

2. **Square of a difference**
$$(x - y)^2 = x^2 - 2xy + y^2.$$

3. **Difference of squares**
$$(x + y)(x - y) = x^2 - y^2.$$

4. **Sum of cubes**
$$x^3 + y^3 = (x + y)(x^2 - xy + y^2).$$

5. **Difference of cubes**
$$x^3 - y^3 = (x - y)(x^2 + xy + y^2).$$

6. **Cube of a sum**
$$(x + y)^3 = x^3 + 3x^2y + 3xy^2 + y^3.$$

7. **Cube of a difference**
$$(x - y)^3 = x^3 - 3x^2y + 3xy^2 - y^3.$$

8. **Binomial theorem**
$$(x + y)^n = \sum_{k=0}^{n} \binom{n}{k} x^{n-k} y^k.$$

9. **Perfect square trinomial**
$$x^2 + 2xy + y^2 = (x + y)^2.$$

10. **Perfect cube trinomial**
$$x^3 + 3x^2y + 3xy^2 + y^3 = (x + y)^3.$$

11. **Difference of higher powers**
$$x^n - y^n = (x - y)\left(x^{n-1} + x^{n-2}y + \cdots + xy^{n-2} + y^{n-1}\right).$$

12. **Sum of higher powers**
$$x^n + y^n = (x + y)\left(x^{n-1} - x^{n-2}y + \cdots - xy^{n-2} + y^{n-1}\right).$$

13. **Higher powers via sum–difference identity**
$$x^n \pm y^n = \frac{1}{2^n} \sum_{k=0}^{n} \left(1 \pm (-1)^k\right) \binom{n}{k} (x - y)^k (x + y)^{n-k}.$$

3.3.6   Inequalities

Solving Inequalities

Solving inequalities involves finding the values of the variable that make the inequality true. The methods are similar to solving equations but require special attention when multiplying or dividing by a negative number (which reverses the inequality sign).

Example: Solve $-2x + 3 > 7$**.**
Solution:

$$-2x + 3 > 7,$$
$$-2x > 4,$$
$$x < -\frac{4}{2},$$
$$x < -2.$$

Answer: All real numbers x such that $x < -2$.

3.3.7   Functions

Definition

A **function** is a relation between a set of inputs and a set of possible outputs, where each input is related to exactly one output. Functions can be represented in various ways, including equations, graphs, tables, or words.

Domain and Range

The **domain** of a function is the set of all possible input values (usually x) for which the function is defined. The **range** is the set of all possible output values (usually $f(x)$).

For the function $f(x) = x^2$:

- **Domain**: all real numbers ($\mathbb{R}$),

- **Range**: all non-negative real numbers ($[0, \infty)$).

3.3.8   Exponential Functions

Definition

An **exponential function** is a function in which the variable appears in the exponent. The general form is

$$f(x) = b^x,$$

where b is a positive constant called the *base*, and x is any real number.

Properties of Exponential Functions

- **Growth or decay**: If $b > 1$, the function models exponential growth. If $0 < b < 1$, it models exponential decay.

- **Addition of exponents**:

$$b^{x+y} = b^x \cdot b^y.$$

- **Power of a power:**

$$(b^x)^y = b^{xy}.$$

- **Zero exponent:**

$$b^0 = 1.$$

- **Negative exponent:**

$$b^{-x} = \frac{1}{b^x}.$$

Natural Exponential Function

When the base b is the mathematical constant e (approximately 2.7182818284), the function is called the **natural exponential function** and is denoted

$$f(x) = e^x.$$

Some of its important properties are:

- The derivative of e^x is e^x:

$$\frac{d}{dx}e^x = e^x.$$

- The function passes through the point $(0, 1)$ since $e^0 = 1$.

- The natural exponential function grows faster than any polynomial as $x \to \infty$.

3.3.9 Logarithms

Definition

A **logarithm** is the inverse operation to exponentiation. For a given base b (with $b > 0$ and $b \neq 1$) and a positive number y, the logarithm of y with base b is denoted by $\log_b y$ and is defined by

$$\log_b y = x \quad \text{if and only if} \quad b^x = y,$$

where $b > 0$, $b \neq 1$, and $y > 0$.

Properties of Logarithms

- **Product rule:**

$$\log_b(xy) = \log_b x + \log_b y.$$

- **Quotient rule:**

$$\log_b\left(\frac{x}{y}\right) = \log_b x - \log_b y.$$

- **Power rule:**

$$\log_b(x^k) = k \log_b x.$$

- **Change of base formula:**

$$\log_b x = \frac{\log_k x}{\log_k b},$$

where k is any positive value different from 1.

- **Inverse property:**

$$b^{\log_b x} = x \quad \text{and} \quad \log_b(b^x) = x.$$

Solving Logarithmic Equations

Example: Solve $\log_2(x) + \log_2(x - 3) = 2.$

　Solution:

　Use the product rule:

$$\log_2\big(x(x - 3)\big) = 2.$$

Simplify inside the logarithm:

$$\log_2(x^2 - 3x) = 2.$$

Convert to exponential form:

$$x^2 - 3x = 2^2 = 4.$$

Bring all terms to one side:

$$x^2 - 3x - 4 = 0.$$

Factor:

$$(x - 4)(x + 1) = 0.$$

So

$$x = 4 \quad \text{or} \quad x = -1.$$

However, in the original equation both x and $x - 3$ must be positive. Thus $x = 4$ is valid, but $x = -1$ is not, since $\log_2(-1)$ is not a real number. (We will learn more about such cases when we study complex analysis.)

3.3.10　Elementary Functions

Elementary functions are the basic building blocks in mathematics. They include polynomials, exponentials, logarithms, trigonometric functions, and their inverses. They are formed through operations such as addition, multiplication, division, and composition. However, not all integrals of elementary functions are themselves elementary.

Categories of Elementary Functions and Examples

Category	Example
Constant function	$f(x) = 5$
Identity function	$f(x) = x$
Power function	$f(x) = x^2$
Polynomial function	$f(x) = 2x^3 - 3x^2 + x - 7$
Rational function	$f(x) = \frac{x}{x-3}$
Exponential function	$f(x) = e^x$
Logarithmic function	$f(x) = \log_{10}(x)$
Trigonometric function	$f(x) = \sin(x)$
Inverse trigonometric function	$f(x) = \arcsin(x)$
Hyperbolic function	$\sinh(x) = \frac{e^x - e^{-x}}{2}$
Inverse hyperbolic function	$\operatorname{arsinh}(x) = \ln\left(x + \sqrt{x^2 + 1}\right)$
Absolute value function	$f(x) = |x|$
Sign function (signum function)	$\operatorname{sgn}(x) = \begin{cases} 1 & \text{if } x > 0, \\ 0 & \text{if } x = 0, \\ -1 & \text{if } x < 0 \end{cases}$
Heaviside step function	$H(x) = \begin{cases} 0 & \text{if } x < 0, \\ 1 & \text{if } x \geq 0 \end{cases}$
Floor function	$f(x) = \lfloor x \rfloor$
Ceiling function	$f(x) = \lceil x \rceil$

Table 3.1: Categories of elementary functions and examples

3.4 Graph

René Descartes

Figure 3.4: René Descartes (1596–1650)

René Descartes (1596–1650) was a French mathematician and philosopher whose work reshaped mathematics by linking algebra with geometry. In *La Géométrie* (1637), he introduced what we now call Cartesian coordinates, showing how curves can be described by equations and how geometric problems can be solved using algebraic methods. This "coordinate geometry" became a cornerstone of modern analysis and laid crucial groundwork for later developments in calculus and mathematical physics.

3.4.1 Coordinate System

Coordinate systems are essential for locating points in a plane. The two primary systems are the Cartesian (rectangular) and polar coordinate systems. Below, we explain each system, illustrate them with the specific point $(3, 4)$, and demonstrate conversions between Cartesian and polar coordinates.

Cartesian Coordinate System (Rectangular Coordinates)

Definition: The Cartesian coordinate system specifies each point uniquely by a pair of numerical coordinates (x, y), which represent the horizontal and vertical distances from two perpendicular axes (the x-axis and y-axis).

Key properties:

- **Axes**: two perpendicular lines; the x-axis (horizontal) and the y-axis (vertical).

- **Origin**: the intersection point of the axes, denoted by $(0, 0)$.

- **Quadrants**: the plane is divided into four quadrants based on the signs of x and y.

Example: Consider the point $(3, 4)$. Here, $x = 3$ and $y = 4$, which places the point in Quadrant I.

Cartesian coordinates: $(3, 4)$.

Polar Coordinate System

Definition: The polar coordinate system represents each point by a distance from the origin, r, and an angle θ measured from the positive x-axis.

Conversion formulas:

$$x = r\cos(\theta), \quad y = r\sin(\theta),$$

$$r = \sqrt{x^2 + y^2}, \quad \theta = \arctan\left(\frac{y}{x}\right).$$

Visual Representation

For illustrative purposes, here is a diagram showing the point $(3, 4)$ in both coordinate systems:

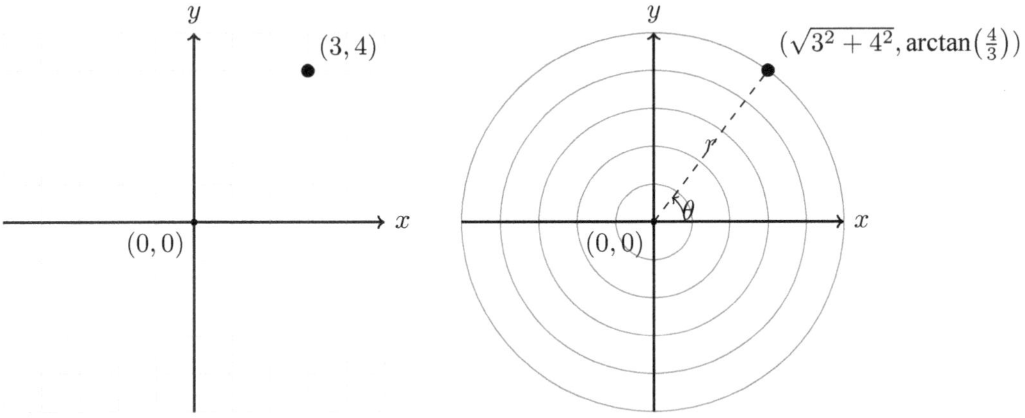

Figure 3.5: Rectangular (left) and polar (right) coordinate systems showing the point $(3, 4)$ in rectangular form and $(\sqrt{3^2 + 4^2}, \arctan(\frac{4}{3}))$ in polar form.

3.4.2 Function Graph

3.4.3 Understanding Growth Rates and Bounding

Growth rates describe how quickly a function's output increases or decreases as the input becomes large. Functions can be categorized by their growth, which is crucial in fields like calculus, number theory, and computer science.

3.4.4 Bounding Functions

Bounding refers to establishing limits within which a function operates. Upper and lower bounds define the maximum and minimum values a function can reach under certain conditions. In algorithm analysis, bounding is used to predict worst-case and best-case scenarios.

3.4.5 Function Charts and Growth Rate

Below are graphs of several functions, arranged from the slowest-growing to the fastest-growing. Understanding these growth rates is essential for analyzing complexity in mathematics, computer science, and related disciplines.

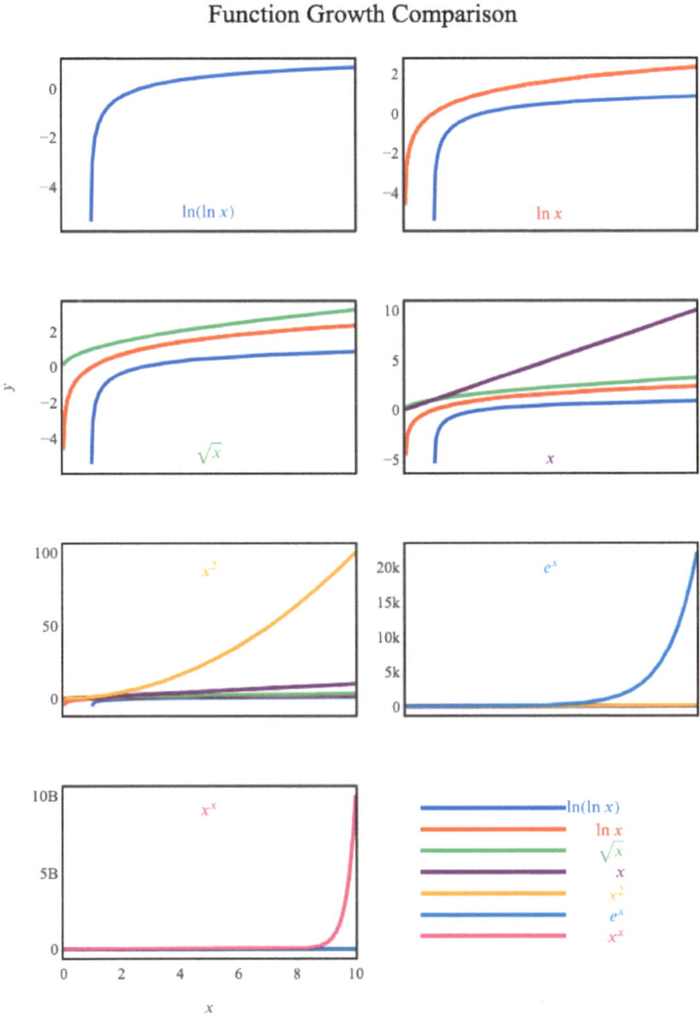

Figure 3.6: Function growth comparison — **Domain**: $[0, 10]$

By comparing these functions visually in this sequence, one can observe how each function grows relative to the others, starting from the very slow growth of $\ln(\ln(x))$ to the explosive growth of x^x.

3.5 Challenge: Gauss Circle Problem

3.5.1 Problem

We now return to the Gauss circle problem introduced earlier.

Imagine a circle of radius r centered at the origin. The goal is to count how many integer points (x, y) lie within or on the boundary of this circle, where

$$x^2 + y^2 \leq r^2.$$

These integer points are called **lattice points**. They appear symmetrically in all four quadrants.

The challenge is to **find a formula** that gives the total number of lattice points $N(r)$ for a given radius r, and then to study how well this can be approximated.

3.5.2 Exact Formula: Gauss Circle Lattice Points

Computing the exact number of lattice points can be done using the Pythagorean theorem and the floor function. One formula for the exact number of lattice points is

$$N(r) = 1 + 4 \sum_{i=1}^{r} \left\lfloor \sqrt{i \cdot (2r - i)} \right\rfloor.$$

Your own formula might look different but should yield the same values. Due to the presence of the floor function, this expression cannot be simplified into a simple closed form and becomes computationally intensive for large r.

We now move on to the approximation aspect, where charting and data-driven methods help us devise and analyze new approximation formulas.

3.5.3 Bounding: Geometric View

Consider the unit–area square with vertices $\{(\pm\frac{1}{2}, \pm\frac{1}{2})\}$. If we translate this square to every lattice point, the squares almost fill the circle's area. Squares touching the boundary form a thin border around the circumference.

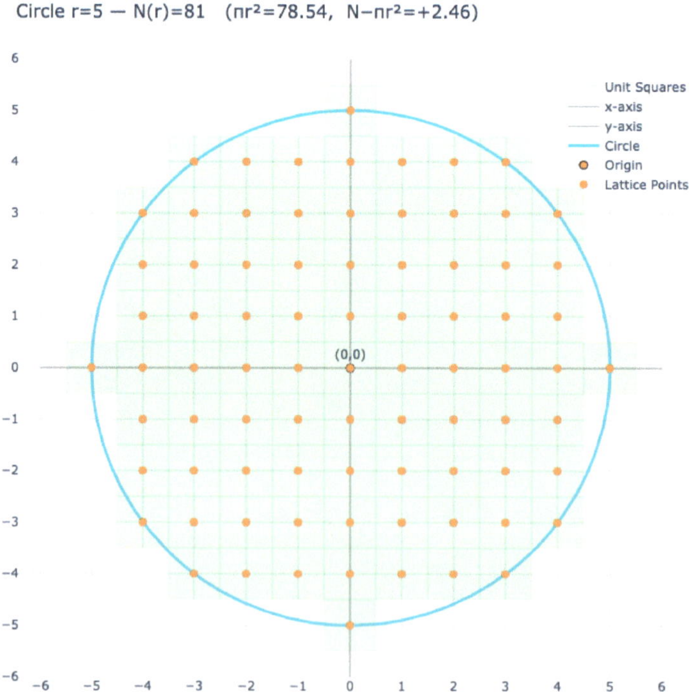

Figure 3.7: Lattice points (orange) and unit squares (green) covering the disk of radius $r = 5$.

The total number of lattice points $N(r)$ corresponds to the total green area. We approximate it by the circle's area plus a small boundary correction:

$$N(r) = A(r) + E(r),$$

where

$$A(r) = \pi r^2, \qquad E(r) = E_{\text{out}}(r) - E_{\text{in}}(r),$$

and

- $E_{\text{out}}(r)$: green area outside the circle,

- $E_{\text{in}}(r)$: missing (void) area inside the circle.

Now let's bound $N(r)$ geometrically. Consider one unit square whose center lies on the circle's boundary. Part of the square lies outside the circle, and part lies inside. If the center is within the circle, the external part contributes a positive correction to $A(r)$; if the center is outside, the missing inner part contributes a negative correction.

Let $s(r, \alpha)$ denote a square whose center lies on the circle of radius r at angle α. The largest deviation of a corner from the circle occurs at angles

$$\alpha = k\frac{\pi}{2} + \frac{\pi}{4},$$

where the square's corner extends the farthest beyond the circle:

$$\sqrt{\left(\tfrac{1}{2}\right)^2 + \left(\tfrac{1}{2}\right)^2} = \tfrac{\sqrt{2}}{2}.$$

Hence, the total green region is bounded between two concentric circles:

$$\pi\left(r - \tfrac{\sqrt{2}}{2}\right)^2 < N(r) < \pi\left(r + \tfrac{\sqrt{2}}{2}\right)^2.$$

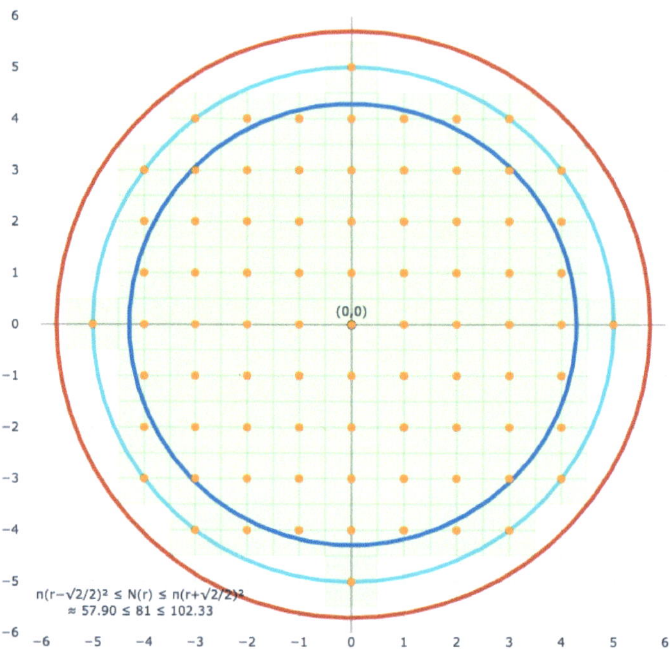

Figure 3.8: Geometric bounds on $N(r)$: the green area lies entirely between the blue and red circles.

Thus, the inner circle with radius $r - \frac{\sqrt{2}}{2}$ lies entirely inside the green region, and no part of the green area extends beyond the outer circle of radius $r + \frac{\sqrt{2}}{2}$. This gives a clean geometric bound for $N(r)$.

However, this $\frac{\sqrt{2}}{2}$ deviation represents only the *maximum* distance from the circumference. So there is not any green point outside of the red cricle.

To estimate a tighter bound, we examine the angles that produce maximum area *per unit of arc*. By Calculus this occurs at the four points $\alpha = k\pi/2$, where each square contributes the largest area outside (or inside) the circle per unit of arc.

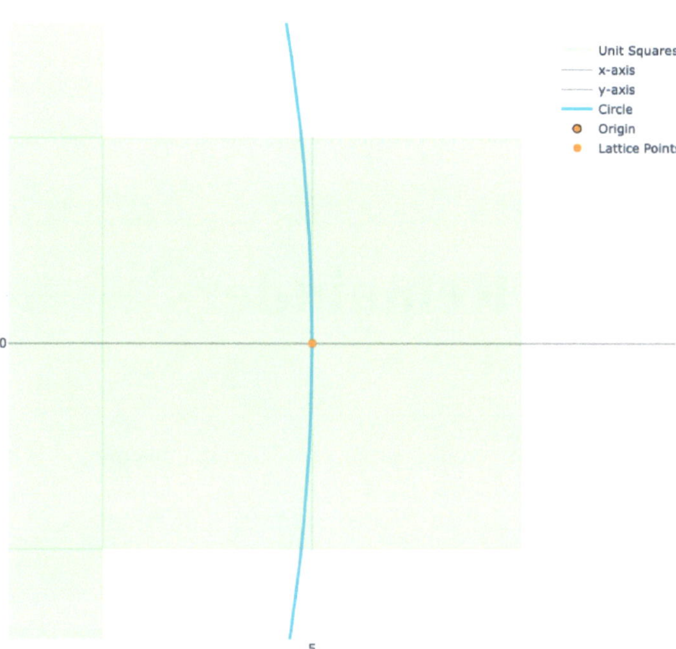

Figure 3.9: Squares aligned with the axes give the maximal area contribution per unit of arc.

At these angles $k\pi/2$, the maximum distance from the circle to the square edge is $\frac{1}{2}$. Using this distance for all angles gives upper and lower error bounds:

$$E_{\text{out}}^{\text{max}} = \pi\left(r + \tfrac{1}{2}\right)^2 - \pi r^2,$$

and symmetrically, the maximum inner deficit is

$$E_{\text{in}}^{\text{max}} = \pi r^2 - \pi\left(r - \tfrac{1}{2}\right)^2.$$

Therefore, a tighter geometric bound is

$$\pi\left(r - \tfrac{1}{2}\right)^2 \;<\; N(r) \;<\; \pi\left(r + \tfrac{1}{2}\right)^2.$$

Even this refinement assumes that all boundary squares are simultaneously in their extreme positions and that every local contribution pushes the error in the same direction. In reality, different parts of the boundary contribute positive and negative errors, and these partially cancel out, so the true error is much smaller than the width of this geometric band. Empirically, the error $E(r) = N(r) - \pi r^2$ oscillates with a slowly growing envelope that is well captured by $\pi\sqrt{r}$. Although this is not the most accurate bound, it provides a a simple, well accurate and practical average approximation.

$$N(r) \approx \pi r^2 + \pi\sqrt{r}.$$

Chapter 4

Exponential Remainders

"I have discovered a truly marvelous proof of this, which this margin is too narrow to contain."
— Pierre de Fermat

Figure 4.1: Pierre de Fermat (1601–1665)

4.1 Preface

"Here was a problem, that I, a ten-year-old, could understand, and I knew from that moment that I would never let it go. I had to solve it."

— Sir Andrew Wiles

4.1.1 Introduction

Calculating exponential remainders is a fundamental concept in mathematics, particularly in number theory and cryptography. It involves finding the remainder when a large exponentiation is divided by a number, specifically computing $m^n \mod c$ for natural numbers m, n, and c.

In this chapter, we will explore the rules and formulas for calculating exponential remainders. Beginning with basic principles, we will progress to more efficient methods, aiming to find a general and efficient way to compute $m^n \mod c$ without unnecessary restrictions.

Throughout this chapter, we will encounter the remainder theorems of Fermat, Euler, and Carmichael, and gain a conceptual understanding of each. We will then apply the strategies of Initial Steps and Periods, learned in the chapter on Remainder Calculation Sequences. By creating datasets and making observations, we will formulate our findings.

While all remainder theorems work well and are efficient under their conditions, they require the initial condition that $\gcd(m, c) = 1$. Our observations show that we can calculate $m^n \mod c$ for all natural numbers m and c, and we will provide a new formulation that does not rely on calculating the GCD.

This chapter is dedicated to two great figures in mathematics—Sunzi and Pierre de Fermat—whose insights into congruences and modular arithmetic underpin the Chinese Remainder Theorem and much of what we build on here.

Pierre de Fermat - Prince of Amateurs

Pierre de Fermat (1601–1665) is one of history's most captivating thinkers because he did world-class mathematics almost as a "side life," while serving as a respected magistrate in southern France—hence his nickname, the "Prince of Amateurs." He rarely published, preferring to challenge friends through letters and margin notes, and that playful, competitive style helped spark whole new directions in mathematics. Fermat had a sharp instinct for patterns in numbers and a talent for turning simple questions into deep ones, and his most famous note—scribbled in the margin of a book, claiming he had found a "truly marvelous" proof that didn't fit—became the legend that haunted mathematicians for centuries. That single line helped fuel generations of inquiry, drawing brilliant minds into a long chase that ultimately connected distant areas of mathematics and culminated in a modern proof in the 1990s. What makes Fermat exciting is not just what he discovered, but *how* he lived his discoveries: quietly, persistently, and with a sense of intellectual adventure that still feels electric today.

Sun Zi (Sunzi) and the Birth of the Chinese Remainder Idea

Sun Zi (Sunzi) is remembered less for a recorded biography and more for a brilliant mathematical voice preserved in the *Sunzi Suanjing* ("Master Sun's Mathematical Classic"), a practical handbook that taught people how to think with numbers through concrete, puzzle-like problems. His most enduring contribution is a famous "remainders" challenge: you are told how a number behaves when counted in different group sizes, and you must reconstruct the hidden total. That single idea—recovering an unknown quantity from several partial constraints—became the seed of what later generations called the Chinese Remainder Theorem, a striking example of how an elegant method can grow out of everyday counting and echo across centuries of mathematics.

Challenge

Find a general formulation to calculate $m^n \mod c$ for $m, n, c \in \mathbb{N}$.

4.1.2 Solution

By analyzing the patterns in $m^n \mod c$ and understanding the concepts of initial steps and periods, we have developed a general method to calculate exponential remainders.

4.1.3 Key Formulas

To calculate $m^n \mod c$:

1. Factorize c into prime factors.

$$c = \prod_{i=1}^{s} p_i^{e_i}$$

2. Calculate the Initial Step Size.

The initial step size is equivalent to the Start Position of the period. Depending on the context, we may use different terminology. For example, if the initial steps have 3 items, the period starts from $n = 3$.

Start Position S:

$$S(c) = \begin{cases} 2 & \text{if } c = 2^3, \\ 1 & \text{if } c = p_1^{e_1}, \\ \max(\text{exponents}) + 1 & \text{otherwise.} \end{cases}$$

3. Calculate the Period.

Period T:

$$T(c) = \begin{cases} 2^{i-2} & \text{if } c = 2^i \text{ and } i \geq 3, \\ (p_1 - 1) \cdot p_1^{e_1 - 1} & \text{if } c = p_1^{e_1}, \\ \text{LCM}\left(T(p_1^{e_1}), T(p_2^{e_2}), \ldots, T(p_s^{e_s})\right) & \text{otherwise.} \end{cases}$$

Note: The period T is equivalent to the Carmichael Lambda Function.

4. Calculate the Remainder (for $n \geq S$):

$$m^n \mod c = m^{((n-S) \mod T) + S} \mod c$$

4.2 Understanding Exponential Remainders

To solve problems like $m^n \mod c$, we need to understand the behavior of the remainders as we raise m to higher powers modulo c.

Consider constructing a table for each c, where:

- **Rows** represent values of m.

- **Columns** represent values of n.

- **Each cell** contains the value of $m^n \mod c$.

By analyzing these tables, we can observe patterns and relationships that help simplify the calculations. Furthermore, we will present different examples from various types to gain insights and advance toward our solution.

4.2.1 Case 1: When c is a Prime Number

Let us begin with the simple case where c is a prime number p. We redefine our problem as computing $m^n \mod p$. We will continue with an example in which $c = 7$; other primes exhibit similar behavior.

Example: $c = 7$

Consider $p = 7$. We construct a table of $m^n \mod 7$ for various values of m and n. Pay close attention to the patterns in the table.

$m^n \mod 7$	m^0	m^1	m^2	m^3	m^4	m^5	m^6	m^7	m^8	m^9	m^{10}	m^{11}	m^{12}	m^{13}	m^{14}	m^{15}
0	1	0	0	0	0	0	0	0	0	0	0	0	0	0	0	0
1	1	1	1	1	1	1	1	1	1	1	1	1	1	1	1	1
2	1	2	4	1	2	4	1	2	4	1	2	4	1	2	4	1
3	1	3	2	6	4	5	1	3	2	6	4	5	1	3	2	6
4	1	4	2	1	4	2	1	4	2	1	4	2	1	4	2	1
5	1	5	4	6	2	3	1	5	4	6	2	3	1	5	4	6
6	1	6	1	6	1	6	1	6	1	6	1	6	1	6	1	6
7	1	0	0	0	0	0	0	0	0	0	0	0	0	0	0	0
8	1	1	1	1	1	1	1	1	1	1	1	1	1	1	1	1
9	1	2	4	1	2	4	1	2	4	1	2	4	1	2	4	1
10	1	3	2	6	4	5	1	3	2	6	4	5	1	3	2	6
11	1	4	2	1	4	2	1	4	2	1	4	2	1	4	2	1
12	1	5	4	6	2	3	1	5	4	6	2	3	1	5	4	6
13	1	6	1	6	1	6	1	6	1	6	1	6	1	6	1	6
14	1	0	0	0	0	0	0	0	0	0	0	0	0	0	0	0
15	1	1	1	1	1	1	1	1	1	1	1	1	1	1	1	1

The interesting pattern is that in the column m^6, except for $m = 7k$, all entries are one. This was discovered by Pierre de Fermat.

Fermat's Little Theorem

Pierre de Fermat noticed these patterns and formulated:

If p is a prime number and m is an integer not divisible by p (i.e., $\gcd(m, p) = 1$), then:

$$m^{p-1} \equiv 1 \mod p.$$

Here are the positions in the table that Fermat's Little Theorem addresses. Looking at our table, for every exponent $n = k(p - 1)$, the remainder cycles back to 1. This periodic property allows us to calculate exponential remainders for large exponents efficiently.

$m^n \bmod 7$	m^0	m^1	m^2	m^3	m^4	m^5	m^6	m^7	m^8	m^9	m^{10}	m^{11}	m^{12}	m^{13}	m^{14}	m^{15}
0	1	0	0	0	0	0	0	0	0	0	0	0	0	0	0	0
1	1	1	1	1	1	1	1	1	1	1	1	1	1	1	1	1
2	1	2	4	1	2	4	1	2	4	1	2	4	1	2	4	1
3	1	3	2	6	4	5	1	3	2	6	4	5	1	3	2	6
4	1	4	2	1	4	2	1	4	2	1	4	2	1	4	2	1
5	1	5	4	6	2	3	1	5	4	6	2	3	1	5	4	6
6	1	6	1	6	1	6	1	6	1	6	1	6	1	6	1	6
7	1	0	0	0	0	0	0	0	0	0	0	0	0	0	0	0
8	1	1	1	1	1	1	1	1	1	1	1	1	1	1	1	1
9	1	2	4	1	2	4	1	2	4	1	2	4	1	2	4	1
10	1	3	2	6	4	5	1	3	2	6	4	5	1	3	2	6
11	1	4	2	1	4	2	1	4	2	1	4	2	1	4	2	1
12	1	5	4	6	2	3	1	5	4	6	2	3	1	5	4	6
13	1	6	1	6	1	6	1	6	1	6	1	6	1	6	1	6
14	1	0	0	0	0	0	0	0	0	0	0	0	0	0	0	0
15	1	1	1	1	1	1	1	1	1	1	1	1	1	1	1	1

Example: Calculating $1234567^{1234567} \bmod 7$ **by Hand**

First, compute $1234567 \bmod 7$ using the **Remainder Calculation Sequence (RCS)** method, as introduced in Chapter 2.

Using RCS for Modulo 7   The RCS for 7 is:

$$\mathrm{RCS}_7 = \{1, 3, 2, -1, -3, -2, \dots\}.$$

$$
\begin{array}{lccccccc}
\text{Digits:} & 1 & 2 & 3 & 4 & 5 & 6 & 7 \\
\mathrm{RCS}_7 \text{ sequence:} & 1 & -2 & -3 & -1 & 2 & 3 & 1
\end{array}
$$

Multiply each digit by the corresponding RCS value and sum them:

$$\text{Sum} = (1 \times 1) + (2 \times -2) + (3 \times -3) + (4 \times -1) + (5 \times 2) + (6 \times 3) + (7 \times 1) = 19.$$

Thus:

$$1234567 \bmod 7 = 19 \bmod 7 = 5.$$

Therefore:

$$1234567^{1234567} \bmod 7 = 5^{1234567} \bmod 7.$$

Since 7 is prime and 5 is not a multiple of 7, Fermat's Little Theorem tells us there is a period of 6 in the exponent. Hence:

$$5^{1234567} \bmod 7 = 5^{(1234567 \bmod 6)} \bmod 7.$$

Next, we need to calculate $1234567 \bmod 6$.

Using RCS for Modulo 6   The RCS for 6 is:

$$\text{RCS}_6 = \{1, -2, -2, \dots\}.$$

Since $-2 \equiv 4 \mod 6$, every -2 in RCS_6 can be replaced by 4 if desired. Thus, we define a custom RCS_6 to minimize computation for this example:

$$\text{RCS}_6 = \{1, -2, -2, 4, -2, 4, -2\}.$$

Align the digits and the RCS sequence, then compute:

$$
\begin{array}{lccccccc}
\text{Digits:} & 1 & 2 & 3 & 4 & 5 & 6 & 7 \\
\text{RCS}_6 \text{ sequence:} & -2 & 4 & -2 & 4 & -2 & -2 & 1
\end{array}
$$

Compute the Sum:

$$\text{Sum} = (1 \times -2) + (2 \times 4) + (3 \times -2) + (4 \times 4) + (5 \times -2) + (6 \times -2) + (7 \times 1).$$

$$\text{Sum} = (-2) + 8 + (-6) + 16 + (-10) + (-12) + 7 = 1.$$

Thus:

$$1234567 \mod 6 = 1.$$

Substituting back:

$$1234567^{1234567} \mod 7 = 5^{1234567} \mod 7 = 5^{(1234567 \mod 6)} \mod 7 = 5^1 \mod 7 = 5.$$

4.2.2   Case 2: When c is a Composite Number

Next, let's explore the case in which c is a composite number. We will examine two examples: $c = 9$ and $c = 12$.

Euler's Theorem

Leonhard Euler extended Fermat's Little Theorem to work with any positive integer c and m coprime to c.

Euler's Theorem. If c is a positive integer and m is an integer such that $\gcd(m, c) = 1$, then:

$$m^{\phi(c)} \equiv 1 \mod c.$$

Here, $\phi(c)$ is Euler's totient function, which counts the positive integers less than c that are coprime to c.

Example One: $c = 9$

Calculate $\phi(9)$:

The numbers less than 9 and coprime to 9 are $\{1, 2, 4, 5, 7, 8\}$, so:

$$\phi(9) = 6.$$

$m^n \mod 9$	m^0	m^1	m^2	m^3	m^4	m^5	m^6	m^7	m^8	m^9	m^{10}	m^{11}	m^{12}	m^{13}	m^{14}	m^{15}
0	1	0	0	0	0	0	0	0	0	0	0	0	0	0	0	0
1	1	1	1	1	1	1	1	1	1	1	1	1	1	1	1	1
2	1	2	4	8	7	5	1	2	4	8	7	5	1	2	4	8
3	1	3	0	0	0	0	0	0	0	0	0	0	0	0	0	0
4	1	4	7	1	4	7	1	4	7	1	4	7	1	4	7	1
5	1	5	7	8	4	2	1	5	7	8	4	2	1	5	7	8
6	1	6	0	0	0	0	0	0	0	0	0	0	0	0	0	0
7	1	7	4	1	7	4	1	7	4	1	7	4	1	7	4	1
8	1	8	1	8	1	8	1	8	1	8	1	8	1	8	1	8
9	1	0	0	0	0	0	0	0	0	0	0	0	0	0	0	0
10	1	1	1	1	1	1	1	1	1	1	1	1	1	1	1	1
11	1	2	4	8	7	5	1	2	4	8	7	5	1	2	4	8
12	1	3	0	0	0	0	0	0	0	0	0	0	0	0	0	0
13	1	4	7	1	4	7	1	4	7	1	4	7	1	4	7	1
14	1	5	7	8	4	2	1	5	7	8	4	2	1	5	7	8
15	1	6	0	0	0	0	0	0	0	0	0	0	0	0	0	0

For all m such that $\gcd(m, 9) = 1$, we have:

$$m^6 \equiv 1 \quad \mod 9.$$

Euler's totient function can be computed using the formula:

$$\phi(n) = n \times \prod_{p \mid n} \left(1 - \frac{1}{p}\right)$$

Example.

$$\phi(9) = 9 \times \left(1 - \frac{1}{3}\right) = 6$$

Example: $c = 12$

Calculate $\phi(12)$:

The numbers less than 12 and coprime to 12 are $\{1, 5, 7, 11\}$, so:

$$\phi(12) = 4.$$

$m^n \mod 12$	m^0	m^1	m^2	m^3	m^4	m^5	m^6	m^7	m^8	m^9	m^{10}	m^{11}	m^{12}	m^{13}	m^{14}	m^{15}
0	1	0	0	0	0	0	0	0	0	0	0	0	0	0	0	0
1	1	1	1	1	1	1	1	1	1	1	1	1	1	1	1	1
2	1	2	4	8	4	8	4	8	4	8	4	8	4	8	4	8
3	1	3	9	3	9	3	9	3	9	3	9	3	9	3	9	3
4	1	4	4	4	4	4	4	4	4	4	4	4	4	4	4	4
5	1	5	1	5	1	5	1	5	1	5	1	5	1	5	1	5
6	1	6	0	6	0	6	0	6	0	6	0	6	0	6	0	6
7	1	7	1	7	1	7	1	7	1	7	1	7	1	7	1	7
8	1	8	4	8	4	8	4	8	4	8	4	8	4	8	4	8
9	1	9	9	9	9	9	9	9	9	9	9	9	9	9	9	9
10	1	10	4	10	4	10	4	10	4	10	4	10	4	10	4	10
11	1	11	1	11	1	11	1	11	1	11	1	11	1	11	1	11
12	1	0	0	0	0	0	0	0	0	0	0	0	0	0	0	0
13	1	1	1	1	1	1	1	1	1	1	1	1	1	1	1	1
14	1	2	4	8	4	8	4	8	4	8	4	8	4	8	4	8
15	1	3	9	3	9	3	9	3	9	3	9	3	9	3	9	3

For all m such that $\gcd(m, 12) = 1$, we have:

$$m^4 \equiv 1 \mod 12.$$

Consequently, the period is:

$$m^{4k} \equiv 1 \mod 12 \quad \text{for } k \in \mathbb{N}.$$

Euler's Remainder System was a significant breakthrough, because it made it possible to calculate the remainder of all exponential numbers, even when c is composite.

However, there are several areas in which Euler's Remainder Theorem can be improved:

1. Calculating $\gcd(m, c)$: This step is time-consuming. 2. Applicability: If $\gcd(m, c) \neq 1$, Euler's Theorem does not directly apply; in such cases, the Chinese Remainder Theorem might help. 3. Efficiency: The Remainder Theorem could be more efficient.

4.3 Analyzing with the Concept of Period

Let's delve deeper into analyzing $m^n \mod c$ without restrictions on m and c.

4.3.1 $m^n \mod c$: Periodicity is Inevitable

It is important to understand the concept of periodicity. If a process has the same conditions and inputs, it will always produce the same output. This is a universal principle, not limited to mathematics.

Understanding $m^n \mod c$ and Its Periodicity

- **Possible Remainders:** When you compute $m^n \mod c$, the possible remainders are in the set $\{0, 1, 2, \ldots, c - 1\}$. This means there are c distinct possible outcomes.

- **Infinite Exponents:** As n increases indefinitely, you continue calculating $m^n \mod c$ for larger and larger exponents.

- **Repetition Guaranteed:** By the Pigeonhole Principle, if you have more exponents than possible remainders (i.e., when $n > c$), at least one remainder must repeat. Thus, there always exists $m^s \equiv m^e \mod c$ for some $s < e$.

Establishing Periodicity   Suppose for two different exponents s and e (with $s < e$), the remainders are equal:

$$m^s \equiv m^e \mod c.$$

- **Synchronizing Remainders:** Once such a repetition occurs, all subsequent remainders follow the same pattern. That is:

$$m^{s+i} \equiv m^{e+i} \mod c \quad \text{for all integers } i \geq 0.$$

Let $T = e - s$:

$$m^{s+i} \equiv m^{s+T+i} \mod c \quad \text{for all integers } i \geq 0.$$

This formula defines the period starting at the point s, with a period T.

4.3.2   The Concept of Initial Steps and Periods

There are two parts to periodicity:

1. **Initial Steps (S):** A finite sequence of remainders before the periodic sequence begins.

2. **Period (T):** The length of the repeating cycle.

The size of the initial steps is equivalent to the Start Position or Start Point of the period. In this chapter, we may also refer to the Initial Step Size as the Start Position, depending on the context.

4.3.3   Observations with $c = 7$

Returning to our earlier example with $c = 7$:

$m^n \mod 7$	m^0	m^1	m^2	m^3	m^4	m^5	m^6	m^7	m^8	m^9	m^{10}	m^{11}	m^{12}	m^{13}	m^{14}	m^{15}
0	1	0	0	0	0	0	0	0	0	0	0	0	0	0	0	0
1	1	1	1	1	1	1	1	1	1	1	1	1	1	1	1	1
2	1	2	4	1	2	4	1	2	4	1	2	4	1	2	4	1
3	1	3	2	6	4	5	1	3	2	6	4	5	1	3	2	6
4	1	4	2	1	4	2	1	4	2	1	4	2	1	4	2	1
5	1	5	4	6	2	3	1	5	4	6	2	3	1	5	4	6
6	1	6	1	6	1	6	1	6	1	6	1	6	1	6	1	6
7	1	0	0	0	0	0	0	0	0	0	0	0	0	0	0	0
8	1	1	1	1	1	1	1	1	1	1	1	1	1	1	1	1
9	1	2	4	1	2	4	1	2	4	1	2	4	1	2	4	1
10	1	3	2	6	4	5	1	3	2	6	4	5	1	3	2	6
11	1	4	2	1	4	2	1	4	2	1	4	2	1	4	2	1
12	1	5	4	6	2	3	1	5	4	6	2	3	1	5	4	6
13	1	6	1	6	1	6	1	6	1	6	1	6	1	6	1	6
14	1	0	0	0	0	0	0	0	0	0	0	0	0	0	0	0
15	1	1	1	1	1	1	1	1	1	1	1	1	1	1	1	1

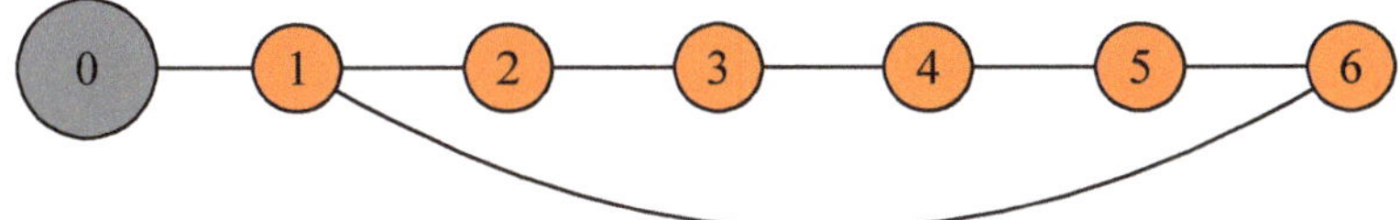

- **Initial Steps (S):** 1 (the sequence starts at $n = 0$).

- **Period (T):** 6 (the remainders repeat every 6 exponents).

By applying Initial Steps and Periods to the columns of the table, we obtain a "periodic box" whose height is always c, but whose width (the period T) can vary.

The period of 6 aligns with Fermat's Little Theorem, where the period is $p - 1$ for a prime p.

4.3.4 Observations with $c = 9$

Now consider $c = 9$:

m^n mod 9	m^0	m^1	m^2	m^3	m^4	m^5	m^6	m^7	m^8	m^9	m^{10}	m^{11}	m^{12}	m^{13}	m^{14}	m^{15}
0	1	0	0	0	0	0	0	0	0	0	0	0	0	0	0	0
1	1	1	1	1	1	1	1	1	1	1	1	1	1	1	1	1
2	1	2	4	8	7	5	1	2	4	8	7	5	1	2	4	8
3	1	3	0	0	0	0	0	0	0	0	0	0	0	0	0	0
4	1	4	7	1	4	7	1	4	7	1	4	7	1	4	7	1
5	1	5	7	8	4	2	1	5	7	8	4	2	1	5	7	8
6	1	6	0	0	0	0	0	0	0	0	0	0	0	0	0	0
7	1	7	4	1	7	4	1	7	4	1	7	4	1	7	4	1
8	1	8	1	8	1	8	1	8	1	8	1	8	1	8	1	8
9	1	0	0	0	0	0	0	0	0	0	0	0	0	0	0	0
10	1	1	1	1	1	1	1	1	1	1	1	1	1	1	1	1
11	1	2	4	8	7	5	1	2	4	8	7	5	1	2	4	8
12	1	3	0	0	0	0	0	0	0	0	0	0	0	0	0	0
13	1	4	7	1	4	7	1	4	7	1	4	7	1	4	7	1
14	1	5	7	8	4	2	1	5	7	8	4	2	1	5	7	8
15	1	6	0	0	0	0	0	0	0	0	0	0	0	0	0	0

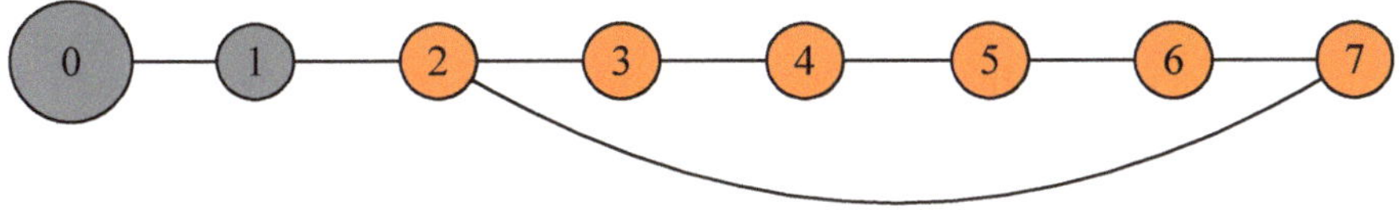

- **Initial Steps (S):** 2

- **Period (T):** 6

Here, the period T is $\phi(9) = 6$, consistent with Euler's Theorem, but the initial steps have grown to $S = 2$. We should investigate the cause.

4.3.5 Observations with $c = 12$

For $c = 12$:

m^n mod 12	m^0	m^1	m^2	m^3	m^4	m^5	m^6	m^7	m^8	m^9	m^{10}	m^{11}	m^{12}	m^{13}	m^{14}	m^{15}
0	1	0	0	0	0	0	0	0	0	0	0	0	0	0	0	0
1	1	1	1	1	1	1	1	1	1	1	1	1	1	1	1	1
2	1	2	4	8	4	8	4	8	4	8	4	8	4	8	4	8
3	1	3	9	3	9	3	9	3	9	3	9	3	9	3	9	3
4	1	4	4	4	4	4	4	4	4	4	4	4	4	4	4	4
5	1	5	1	5	1	5	1	5	1	5	1	5	1	5	1	5
6	1	6	0	6	0	6	0	6	0	6	0	6	0	6	0	6
7	1	7	1	7	1	7	1	7	1	7	1	7	1	7	1	7
8	1	8	4	8	4	8	4	8	4	8	4	8	4	8	4	8
9	1	9	9	9	9	9	9	9	9	9	9	9	9	9	9	9
10	1	10	4	10	4	10	4	10	4	10	4	10	4	10	4	10
11	1	11	1	11	1	11	1	11	1	11	1	11	1	11	1	11
12	1	0	0	0	0	0	0	0	0	0	0	0	0	0	0	0
13	1	1	1	1	1	1	1	1	1	1	1	1	1	1	1	1
14	1	2	4	8	4	8	4	8	4	8	4	8	4	8	4	8
15	1	3	9	3	9	3	9	3	9	3	9	3	9	3	9	3

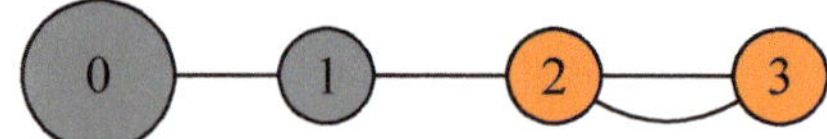

- **Initial Steps (S):** 2

- **Period (T):** 2

We have already seen that $\phi(12) = 4$, but here the period $T = 2$ is shorter than $\phi(12)$. This shows that the actual period can be shorter than Euler's $\phi(c)$.

4.3.6 Understanding the Patterns

After the initial steps, the sequence of remainders enters a cycle of length T. Within that cycle, the remainders repeat in the same order indefinitely.

4.3.7 Data Analysis

Let us prepare a dataset of values of c, their factorizations, and observed values of S and T. By examining this data, we aim to identify patterns and formulate methods for calculating S and T.

c	Factors	S	T
2	2^1	1	1
3	3^1	1	2
4	2^2	1	2
5	5^1	1	4
6	$2^1 \times 3^1$	2	2
7	7^1	1	6
8	2^3	2	2
9	3^2	1	6
10	$2^1 \times 5^1$	2	4
11	11^1	1	10
12	$2^2 \times 3^1$	3	2
13	13^1	1	12
14	$2^1 \times 7^1$	2	6
15	$3^1 \times 5^1$	2	4
16	2^4	1	4
17	17^1	1	16
18	$2^1 \times 3^2$	3	6
19	19^1	1	18
20	$2^2 \times 5^1$	3	4
21	$3^1 \times 7^1$	2	6
22	$2^1 \times 11^1$	2	10
23	23^1	1	22
24	$2^3 \times 3^1$	4	2
25	5^2	1	20
26	$2^1 \times 13^1$	2	12
27	3^3	1	18
28	$2^2 \times 7^1$	3	6
29	29^1	1	28
30	$2^1 \times 3^1 \times 5^1$	2	4
31	31^1	1	30
32	2^5	1	8
33	$3^1 \times 11^1$	2	10
34	$2^1 \times 17^1$	2	16

c	Factors	S	T
35	$5^1 \times 7^1$	2	12
36	$2^2 \times 3^2$	3	6
37	37^1	1	36
38	$2^1 \times 19^1$	2	18
39	$3^1 \times 13^1$	2	12
40	$2^3 \times 5^1$	4	4
41	41^1	1	40
42	$2^1 \times 3^1 \times 7^1$	2	6
43	43^1	1	42
44	$2^2 \times 11^1$	3	10
45	$3^2 \times 5^1$	3	12
46	$2^1 \times 23^1$	2	22
47	47^1	1	46
48	$2^4 \times 3^1$	5	4
49	7^2	1	42
50	$2^1 \times 5^2$	3	20
51	$3^1 \times 17^1$	2	16
52	$2^2 \times 13^1$	3	12
53	53^1	1	52
54	$2^1 \times 3^3$	4	18
55	$5^1 \times 11^1$	2	20
56	$2^3 \times 7^1$	4	6
57	$3^1 \times 19^1$	2	18
58	$2^1 \times 29^1$	2	28
59	59^1	1	58
60	$2^2 \times 3^1 \times 5^1$	3	4
61	61^1	1	60
62	$2^1 \times 31^1$	2	30
63	$3^2 \times 7^1$	3	6
64	2^6	1	16
65	$5^1 \times 13^1$	2	12
66	$2^1 \times 3^1 \times 11^1$	2	10
67	67^1	1	66

c	Factors	S	T
68	$2^2 \times 17^1$	3	16
69	$3^1 \times 23^1$	2	22
70	$2^1 \times 5^1 \times 7^1$	2	12
71	71^1	1	72
72	$2^3 \times 3^2$	4	6
73	73^1	1	72
74	$2^1 \times 37^1$	2	36
75	$3^1 \times 5^2$	3	20
76	$2^2 \times 19^1$	3	18
77	$7^1 \times 11^1$	2	30
78	$2^1 \times 3^1 \times 13^1$	2	12
79	79^1	1	78
80	$2^4 \times 5^1$	5	4
81	3^4	1	54
82	$2^1 \times 41^1$	2	40
83	83^1	1	82
84	$2^2 \times 3^1 \times 7^1$	3	6
85	$5^1 \times 17^1$	2	16
86	$2^1 \times 43^1$	2	42
87	$3^1 \times 29^1$	2	28
88	$2^3 \times 11^1$	4	10
89	89^1	1	88
90	$2^1 \times 3^2 \times 5^1$	3	12
91	$7^1 \times 13^1$	2	12
92	$2^2 \times 23^1$	3	22
93	$3^1 \times 31^1$	2	30
94	$2^1 \times 47^1$	2	46
95	$5^1 \times 19^1$	2	36
96	$2^5 \times 3^1$	6	8
97	97^1	1	96
98	$2^1 \times 7^2$	3	42
99	$3^2 \times 11^1$	3	30
100	$2^2 \times 5^2$	3	20

4.3.8 Determining the Start Position S

The start position S represents the number of initial steps before the cycle begins. From observation:

- For all prime numbers and powers of prime numbers (except 2^3), $S = 1$.

- For $c = 8 = 2^3$, $S = 2$.

- For composite numbers with multiple distinct prime factors, $S = \max(\text{exponents}) + 1$.

$c = 8 = 2^3$ is a particularly interesting case.

Formula for Start Position S:

$$
S(c) = \begin{cases} 2 & \text{if } c = 2^3, \\ 1 & \text{if } c \text{ has only one distinct prime factor,} \\ \max(\text{exponents}) + 1 & \text{otherwise.} \end{cases}
$$

4.3.9   Determining the Period T

Period for Powers of 2

For $c = 2^i$:

$$
T(2^i) = \begin{cases} 2^{i-2}, & \text{if } i \geq 3, \\ 2^{i-1}, & \text{if } i < 3. \end{cases}
$$

Period for Powers of Odd Primes

For $c = p^e$, where p is an odd prime:

$$
T(p^e) = (p - 1)\, p^{e-1}.
$$

Period for Composite Numbers with Multiple Distinct Primes

If

$$
c = \prod_{i=1}^{s} p_i^{e_i},
$$

then

$$
T(c) = \text{LCM}\Big(T(p_1^{e_1}),\ T(p_2^{e_2}),\ \ldots,\ T(p_s^{e_s}) \Big).
$$

This important observation was first made by the American mathematician Robert D. Carmichael, leading to the Carmichael Remainder Theorem, an efficient exponential remainder theorem.

The **Carmichael function** $\lambda(c)$ of a positive integer c is defined as the smallest positive integer n such that:

$$
m^n \equiv 1 \mod c
$$

for all integers m coprime to c. This function generalizes Euler's Theorem and provides the exact period T for modular exponentiation. Thus, the period $T(c)$ we have been calculating is essentially $\lambda(c)$, the Carmichael function of c.

4.3.10   Summary of Period Determination

$$
T(c) = \begin{cases} 2^{i-2} & \text{if } c = 2^i \text{ and } i \geq 3, \\ (p_1 - 1)\, p_1^{e_1 - 1} & \text{if } c = p_1^{e_1}, \\ \text{LCM}(T(p_1^{e_1}), \ldots, T(p_s^{e_s})) & \text{if } c \text{ has multiple distinct prime factors.} \end{cases}
$$

Here, $c = \prod_{i=1}^{s} p_i^{e_i}$, and LCM denotes the least common multiple.

Examples

Calculate the Start Position and Period for the numbers 12 and 66.

Example 1: $c = 12$

Factorization:
$$12 = 2^2 \times 3^1.$$

- **Start Position:**
$$S(12) = \max(2, 1) + 1 = 3.$$

- **Period:**
$$T(12) = \mathrm{LCM}\big(T(2^2), T(3^1)\big) = \mathrm{LCM}(2, 2) = 2.$$

Example 2: $c = 66$

Factorization:
$$66 = 2^1 \times 3^1 \times 11^1.$$

- **Start Position:**
$$S(66) = \max(1, 1, 1) + 1 = 2.$$

- **Period:**
$$T(66) = \mathrm{LCM}\big(T(2^1), T(3^1), T(11^1)\big) = \mathrm{LCM}(1, 2, 10) = 10.$$

4.3.11 Improving and Formulating the Remainder Calculation

In all of Fermat's, Euler's, and Carmichael's remainder theorems for $m^n \mod c$, calculating $\gcd(m, c)$ is required, which is time-consuming. However, by using our approach, we can address this issue. We have observed that there is a "periodic box" that repeats indefinitely. It does not matter whether the items in the box have $\gcd(m, c) = 1$; the entire box repeats as is.

Here is the formula that maps numbers to the first period region:

For $n \geq S$:
$$m^n \mod c = m^{\left((n-S) \mod T\right)+S} \mod c.$$

This formula maps every number to the first period after the initial steps. However, it may not be the most efficient in terms of the exponent. In the previous formulation, all numbers map to the first period, but if the exponent maps to the initial steps and $\gcd(m, c) = 1$, then the value in the initial steps is also correct and does not need adjustment within the first cycle.

The formulas certainly produce an equivalent number in the period after S. Their advantage is that they do not require calculating the GCD. Nevertheless, in cases where $(n \mod T) < S$, we have the option to compute $\gcd(m, c)$. If $\gcd(m, c) = 1$, then we do not need to adjust to the first period; we can directly calculate the remainder with a more optimized exponent, in exchange for computing $\gcd(m, c)$.

If $n \mod T < S$ and $\gcd(m, c) = 1$, then:
$$m^n \mod c = m^{n \mod T} \mod c.$$

Thus, here is the optimized and efficient formula for the exponent:

- If $(n \bmod T) \geq S$, then:

$$m^n \bmod c = m^{(n \bmod T)} \bmod c.$$

- Else if $(n \bmod T) < S$ and $\gcd(m, c) = 1$, then:

$$m^n \bmod c = m^{(n \bmod T)} \bmod c.$$

- Otherwise, add the period T to move within the first cycle range:

$$m^n \bmod c = m^{\left((n \bmod T)+T\right)} \bmod c.$$

4.3.12   Examples of Calculating $m^n \bmod c$

Example 4: $m^{1013} \bmod 66$

Compute $m^{1013} \bmod 66$ for all positive integers m.

- **Determine S and T:**
$$S(66) = 2, \quad T(66) = 10.$$

- **Compute the Remainder:**

$$m^{1013} \bmod 66 = m^{\left((1013-2) \bmod 10\right)+2} \bmod 66 = m^3 \bmod 66,$$

for all positive integers m.

4.4   Summary and Conclusion

By analyzing the patterns in $m^n \bmod c$ and applying the concepts of initial steps and periods, we have developed a general method to calculate exponential remainders.

4.4.1   Key Formulas

Start Position S:

$$S(c) = \begin{cases} 2 & \text{if } c = 2^3, \\ 1 & \text{if } c \text{ has only one distinct prime factor,} \\ \max(\text{exponents}) + 1 & \text{otherwise.} \end{cases}$$

Period T:

$$T(c) = \begin{cases} 2^{i-2} & \text{if } c = 2^i \text{ and } i \geq 3, \\ (p_1 - 1)\, p_1^{e_1 - 1} & \text{if } c = p_1^{e_1}, \\ \mathrm{LCM}\left(T(p_1^{e_1}), T(p_2^{e_2}), \ldots, T(p_s^{e_s})\right) & \text{otherwise.} \end{cases}$$

$$m^n \bmod c = m^{\left((n-S) \bmod T\right)+S} \bmod c.$$

In cases where $(n \mod T) < S$ and $\gcd(m, c) = 1$, the formula can be further optimized:

$$m^n \mod c = m^{(n \mod T)} \mod c.$$

This version is used when $(n \mod T)$ lies in the initial-steps region. Compared with the previous formula, it trades the extra work of computing $\gcd(m, c)$ for a smaller exponent: if $\gcd(m, c) = 1$, there is no need to shift the exponent into the first periodic region.

By applying these formulas, we can efficiently compute $m^n \mod c$ for any natural numbers m, n, and c.

Chapter 5

Periods and Primes

"Discovery is relative to the point of view."

— *Robert D. Carmichael,* The Logic of Discovery

Figure 5.1: Robert Daniel Carmichael (1879–1967)

5.1 Preface

This chapter is dedicated to Franz Mertens, Édouard Lucas, and Robert Daniel Carmichael—whose ideas on prime-density products, Lucas sequences, and Carmichael numbers continue to shape how we test, approximate, and understand prime structure.

Robert D. Carmichael

Robert Daniel Carmichael was a distinguished American mathematician whose contributions have left a lasting impact on the field of number theory. Best known for identifying Carmichael numbers— composite numbers that satisfy Fermat's Little Theorem for all integers relatively prime to them—his work has been pivotal in advancing our understanding of prime number distributions and cryptographic algorithms. Carmichael's keen insights and rigorous approach not only deepened mathematical theory but also provided essential tools for modern computational applications.

Beyond his research achievements, Carmichael was a respected educator and leader within the mathematical community. He served as president of the American Mathematical Society, where he advocated for the expansion of mathematical research and education in the United States. His dedication to mentoring young mathematicians and fostering collaborative scholarly environments helped shape the next generation of mathematical thought. Carmichael's legacy endures through his substantial contributions to mathematics and his unwavering commitment to the advancement of the discipline.

5.1.1 Introduction

In the previous chapter, we focused on understanding the period T in exponential remainders, specifically $m^n \bmod c$. We observed that Pierre de Fermat identified the period for prime numbers as $T(c) = c - 1$. Euler later generalized this period to all prime and composite numbers using the Euler totient function $T(c) = \phi(c)$. However, this does not always yield the fundamental period; sometimes it results in multiples of the fundamental period. Robert Carmichael noticed this discrepancy and further generalized the formula to determine the fundamental period, denoted by $T(c) = \gamma(c)$.

$$T(c) = \gamma(c) = \begin{cases} 2^{i-2} & \text{if } c = 2^i \text{ and } i \geq 3, \\ (p - 1) \cdot p^{e-1} & \text{if } c = p^e, \\ \mathrm{LCM}\big(T(p_1^{e_1}), T(p_2^{e_2}), \ldots, T(p_s^{e_s})\big) & \text{otherwise,} \end{cases}$$

where $c = \prod_{i=1}^{s} p_i^{e_i}$.

When calculating exponential remainders, directly computing large exponents is impractical due to their size, making storage and computation infeasible. However, leveraging the concept of periodicity allows us to compute exponential remainders of large numbers. In this chapter, we delve deeper into the period and its properties, focusing on the relationship between the period and prime numbers.

In the last chapter, we constructed tables to calculate $m^n \bmod c$, where m was listed in the rows, n represented the exponents, and $m^n \bmod c$ filled the corresponding cells.

After several initial steps S, for all m, we have:

$$m^{n+T} \bmod c = m^n \bmod c.$$

Please note that when we refer to the "Period", we mean the *fundamental* period. We also define S and T as follows: after the initial steps, the period is valid for all numbers. We have also observed that if $\gcd(m, c) = 1$,

$$m^n \mod c = m^{n \bmod T(c)} \mod c$$

is always true, even if $n \mod T(c)$ falls within the initial steps.

In this chapter, we present significant relationships within the period column $m^{T(c)} \mod c$ and the factors of c. We explore how these relationships can help determine whether a number is prime. We will analyze the Fermat Test and focus on Carmichael numbers to understand why they are special. Additionally, we present a challenge to solidify these concepts, and finally, we analyze Mersenne numbers.

There is outstanding knowledge in the column of the period $m^{T(c)}$. In summary:

- If $\gcd(m, c) = 1$, then $m^{T(c)} \equiv 1 \pmod{c}$.

- If $\gcd(m, c) \neq 1$, then $m^{T(c)} \not\equiv 1 \pmod{c}$; in many examples the remainder behaves like a non-trivial multiple of m, revealing information about shared factors of m and c.

5.1.2   Examples of Period Columns

In this section, our goal is to understand what is happening in the period column in exponential congruences. Here are examples that illustrate the period column for both prime and composite numbers, demonstrating how these periods help determine the factors of the number.

Example $c = 6$
Carmichael lambda function: $\lambda(6) = 2$.
Primes m from 2 to $\sqrt{6} \approx 2$:

$$2$$

Calculating $m^T \mod c$ for each prime m:

$$2^2 \mod 6 = 4.$$

Result: 6 is a composite number, and $2 \mid 6$.

Example $c = 9$
Factorization: $9 = 3^2$.

$$T(3^2) = (3 - 1) \cdot 3^{2-1} = 2 \cdot 3 = 6.$$

Carmichael lambda function: $\lambda(9) = 6$.
Primes m from 2 to $\sqrt{9} = 3$:

$$2, 3$$

Calculating $m^T \mod c$ for each prime m:

$$2^6 \mod 9 = 1,$$
$$3^6 \mod 9 = 0.$$

Result: 9 is a composite number, and $3 \mid 9$.

Example $c = 13$
Carmichael lambda function: $\lambda(13) = 12$.
Primes m from 2 to $\sqrt{13} \approx 3$:
$$2, 3$$

Calculating $m^T \mod c$ for each prime m:
$$2^{12} \mod 13 = 1,$$
$$3^{12} \mod 13 = 1.$$

Result: 13 is a prime number.

Example $c = 20$
Carmichael lambda function: $\lambda(20) = 4$.
Primes m from 2 to $\sqrt{20} \approx 4$:
$$2, 3$$

Calculating $m^T \mod c$ for each prime m:
$$2^4 \mod 20 = 16,$$
$$3^4 \mod 20 = 1.$$

Result: 20 is a composite number, and $2 \mid 20$.

Example $c = 101$
Carmichael lambda function: $\lambda(101) = 100$.
Primes m from 2 to $\sqrt{101} \approx 10$:
$$2, 3, 5, 7$$

Calculating $m^T \mod c$ for each prime m:
$$2^{100} \mod 101 = 1,$$
$$3^{100} \mod 101 = 1,$$
$$5^{100} \mod 101 = 1,$$
$$7^{100} \mod 101 = 1.$$

Result: 101 is a prime number.

Example $c = 1001$
Carmichael lambda function: $\lambda(1001) = 60$.
Primes m from 2 to $\sqrt{1001} \approx 31$:
$$2, 3, 5, 7, 11, 13, 17, 19, 23, 29, 31$$

Calculating $m^T \mod c$ for each prime m:

$$2^{60} \mod 1001 = 1, \qquad 17^{60} \mod 1001 = 1,$$
$$3^{60} \mod 1001 = 1, \qquad 19^{60} \mod 1001 = 1,$$
$$5^{60} \mod 1001 = 1, \qquad 23^{60} \mod 1001 = 1,$$
$$7^{60} \mod 1001 = 287, \qquad 29^{60} \mod 1001 = 1,$$
$$11^{60} \mod 1001 = 638, \qquad 31^{60} \mod 1001 = 1,$$
$$13^{60} \mod 1001 = 78,$$

Result: 1001 is a composite number, and $7 \mid 1001$, $11 \mid 1001$, and $13 \mid 1001$.

Example $c = 10001$
Carmichael lambda function: $\lambda(10001) = 1224$.
Primes m **from** 2 **to** $\sqrt{10001} \approx 100$**:**

2, 3, 5, 7, 11, 13, 17, 19, 23, 29, 31, 37, 41, 43, 47, 53, 59, 61, 67, 71, 73, 79, 83, 89, 97

$$2^{1224} \bmod 10001 = 1, \qquad 43^{1224} \bmod 10001 = 1,$$
$$3^{1224} \bmod 10001 = 1, \qquad 47^{1224} \bmod 10001 = 1,$$
$$5^{1224} \bmod 10001 = 1, \qquad 53^{1224} \bmod 10001 = 1,$$
$$7^{1224} \bmod 10001 = 1, \qquad 59^{1224} \bmod 10001 = 1,$$
$$11^{1224} \bmod 10001 = 1, \qquad 61^{1224} \bmod 10001 = 1,$$
$$13^{1224} \bmod 10001 = 1, \qquad 67^{1224} \bmod 10001 = 1,$$
$$17^{1224} \bmod 10001 = 1, \qquad 71^{1224} \bmod 10001 = 1,$$
$$19^{1224} \bmod 10001 = 1, \qquad 73^{1224} \bmod 10001 = 8906,$$
$$23^{1224} \bmod 10001 = 1, \qquad 79^{1224} \bmod 10001 = 1,$$
$$29^{1224} \bmod 10001 = 1, \qquad 83^{1224} \bmod 10001 = 1,$$
$$31^{1224} \bmod 10001 = 1, \qquad 89^{1224} \bmod 10001 = 1,$$
$$37^{1224} \bmod 10001 = 1, \qquad 97^{1224} \bmod 10001 = 1,$$
$$41^{1224} \bmod 10001 = 1,$$

Result: 10001 is a composite number, and $73 \mid 10001$.

It's easy and great: by simply examining the column $m^{T(c)}$, we can determine whether a number is prime or composite and even identify factors of the numbers. So to find out if a candidate number c is prime or composite, we look at the results for $m^{T(c)} \bmod c$.

The next challenge is to find the Period $T(c)$. In applications, unfortunately, we cannot use the Carmichael function directly, because it requires you to have already factorized c. If we know the factors, then we already know whether c is prime or composite!

The remaining way is the concept of the period explained in the chapter on Exponential Remainders. There we explain how to find the period, but these methods need to check all conditions and can still be expensive.

Mathematicians, however, look for more efficient ways to know if a number is prime or not; and the shortcut is finding a contradiction. Consider that we want to know if c is prime or composite. We assume c is prime. Prime numbers have unique properties. If we find any conflict with those properties, then c is composite; if c satisfies all the required properties, then c is prime.

We will discuss the Fermat Test. We will analyze it and show its power. There are special composite numbers that can pass many of the Fermat tests. We will encounter Carmichael numbers, analyzing why they are special. These special numbers are the subject of the next chapter.

5.2 Challenge

Here are three numbers: 541, 561, and 703.

One of them is a prime number, another is a Carmichael number, and the third is an ordinary composite number. You are tasked with proving the classification of each.

5.3 Periods and Primes

We know that m^n mod c behaves periodically in both the base m and the exponent n. The reason is that, after reducing modulo c, we have only finitely many states or outputs $\{0, 1, 2, \ldots, c-1\}$, and from each state the next state is determined uniquely. So the system is eventually periodic in both directions.

For a fixed base m with $\gcd(m, c) = 1$, the sequence in the exponent

$$n \longmapsto m^n \bmod c$$

has exponent period $T(c) = \gamma(c)$. For a fixed exponent n, the function in the base

$$m \longmapsto m^n \bmod c$$

is periodic with period c, since $m \equiv m + c \pmod{c}$.

5.3.1 Period in exponential congruent

In the last chapter, we determined the period $T(c)$ by calculating exponential remainders using tables. After several initial steps S, for all m, we have:

$$m^{n+T} \bmod c = m^n \bmod c.$$

Fermat was the first to notice the period in exponents. He states that for every prime p,

$$m^{p-1} \bmod p = 1 \quad \text{for all } m \in \{1, 2, \ldots, p-1\}.$$

Let's consider an example with $p = 7$. We create a table for m^n mod 7. Understanding the rows and columns of this table is crucial, and we will use this concept throughout the chapter. For simplification and clarity, we use elements of the table instead of only abstract conditions.

As Fermat discovered, the exponent period of prime numbers p is $T(p) = p - 1$.

m^n mod 7	m^0	m^1	m^2	m^3	m^4	m^5	m^6	m^7	m^8	m^9	m^{10}	m^{11}	m^{12}	m^{13}	m^{14}	m^{15}
0	1	0	0	0	0	0	0	0	0	0	0	0	0	0	0	0
1	1	1	1	1	1	1	1	1	1	1	1	1	1	1	1	1
2	1	2	4	1	2	4	1	2	4	1	2	4	1	2	4	1
3	1	3	2	6	4	5	1	3	2	6	4	5	1	3	2	6
4	1	4	2	1	4	2	1	4	2	1	4	2	1	4	2	1
5	1	5	4	6	2	3	1	5	4	6	2	3	1	5	4	6
6	1	6	1	6	1	6	1	6	1	6	1	6	1	6	1	6
7	1	0	0	0	0	0	0	0	0	0	0	0	0	0	0	0
8	1	1	1	1	1	1	1	1	1	1	1	1	1	1	1	1
9	1	2	4	1	2	4	1	2	4	1	2	4	1	2	4	1
10	1	3	2	6	4	5	1	3	2	6	4	5	1	3	2	6
11	1	4	2	1	4	2	1	4	2	1	4	2	1	4	2	1
12	1	5	4	6	2	3	1	5	4	6	2	3	1	5	4	6
13	1	6	1	6	1	6	1	6	1	6	1	6	1	6	1	6
14	1	0	0	0	0	0	0	0	0	0	0	0	0	0	0	0
15	1	1	1	1	1	1	1	1	1	1	1	1	1	1	1	1

The period is $T(7) = 6$. The reason we expect the column m^6 to contain all ones is that we know $m^0 \bmod 7 = 1$, and with the period $T = 6$, the column m^6 repeats the values from m^0. Therefore, ones appear at $m^6, m^{12}, m^{18}, \ldots$.

In other words, Fermat demonstrated that the column preceding prime numbers consists entirely of ones from $m = 1$ to $m = p - 1$ when the modulus is a prime p. The condition holds true, and the converse is also true: there is no composite number that produces all ones in the period column $m^{p-1} \bmod p$ for $m = 1, \ldots, p - 1$.

A number p is prime if and only if, for all $m = 1$ to $p - 1$:

$$m^{p-1} \bmod p = 1.$$

Also, we have this useful statement: a number c is composite if and only if there exists some $m \in \{1, \ldots, c - 1\}$ such that

$$m^{c-1} \bmod c \neq 1.$$

To optimize the base options $m = 0$ to $(c - 1)$, note two special cases in our base period: $m = 0$ and $m = 1$:

$$0^n \bmod c = 0, \qquad 1^n \bmod c = 1.$$

These provide no information about the primality of c. Also, from what we learned about the period column, prime bases under $\sqrt{c}$ already suffice to identify the factors of the number and its status. Hence we use m as prime numbers under $\sqrt{c}$.

Based on what we have seen, these statements are equivalent:

1. A number c is a prime number.

2. The period $T(c) = c - 1$.

3. For all primes p from 2 to $\sqrt{c}$:
$$p^{c-1} \bmod c = 1.$$

Alternatively we can describe the composite case:

4. A number c is composite.

5. The period $T(c) < c - 1$.

6. If you find any prime $p \leq \sqrt{c}$ such that $p^{c-1} \bmod c \neq 1$, then c is composite.

For calculating $T(c)$, we can address it using the period concept in the chapter on Exponential Remainders. However, it may not be feasible for large numbers. Specifically, the period of a prime p is $p - 1$, so finding it for a huge p is often not practical. Also, using the Carmichael function is not possible as it requires factorizing c. So we go with the third option, known as the Fermat tests.

5.4   Fermat Test

To check whether a number c is composite, we begin by assuming c is prime. Then, for every prime m from 2 up to $\sqrt{c}$, we expect

$$m^{c-1} \bmod c = 1.$$

If any instance violates this condition, the assumption is proven false, and c must be composite. This approach is rooted in a proof by contradiction. In other words, if for every prime m from 2 to $\sqrt{c}$ we have

$$m^{c-1} \mod c = 1,$$

then c is a very strong prime candidate. However, for large c, testing all such primes is not feasible, so we typically limit the bases and apply these tests.

5.4.1 Accuracy of Fermat Tests (Empirically)

Fix a prime base b. The base-b Fermat test declares an odd integer $c > 1$ to be "probably prime" if

$$b^{c-1} \equiv 1 \pmod{c}.$$

Every prime c with $\gcd(b, c) = 1$ passes this test, but some composite numbers do as well; these are the *Fermat pseudoprimes* to base b.

We computed all pseudoprimes up to 2×10^7 for the bases

$$b \in \{2, 3, 5, 7, 11, 13, 17\}.$$

For each base b let $P_b(c)$ be the number of composite $n \le c$ that pass the base-b test. The following plot shows the functions $P_b(c)$ together with the curve

$$z(c) = \tfrac{1}{4}\sqrt{c}.$$

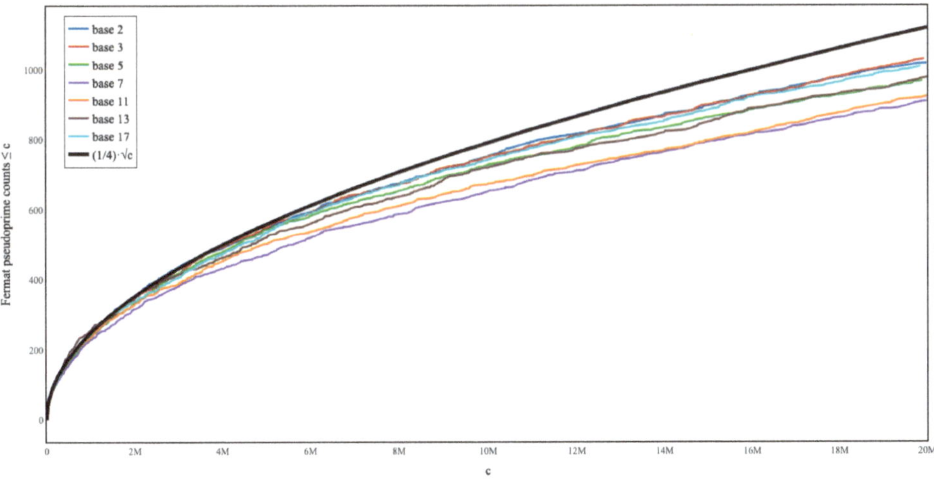

Figure 5.2: Fermat pseudoprime counts up to 2×10^7 for bases $b \in \{2, 3, 5, 7, 11, 13, 17\}$; the black curve is $z(c) = \tfrac{1}{4}\sqrt{c}$.

Empirically, for all these bases and for all $c \le 2 \times 10^7$ we observe

$$P_b(c) \le \tfrac{1}{4}\sqrt{c}.$$

Among these bases, $b = 7$ has the smallest pseudoprime count on this range. The further a base b moves away from 7, the higher its curve $P_b(c)$ lies, and the worse the Fermat test performs for that base (within this experimental range).

Let $\pi(c)$ be the number of primes $\leq c$. The base-b *Fermat candidates* up to c are

$$C_b(c) := \pi(c) + P_b(c),$$

so the (empirical) *accuracy* of the base-b test is

$$\text{Accuracy}_b(c) := \frac{\pi(c)}{C_b(c)} = \frac{\pi(c)}{\pi(c) + P_b(c)}.$$

Using the bound $P_b(c) \leq \frac{1}{4}\sqrt{c}$ for $b \in \{2, 3, 5, 7, 11, 13, 17\}$ we obtain the lower bound

$$\text{Accuracy}_b(c) \geq \frac{\pi(c)}{\pi(c) + \frac{1}{4}\sqrt{c}}.$$

By the Prime Number Theorem, $\pi(c) \sim c/\ln c$, so

$$\text{Accuracy}_b(c) \geq 1 - \Theta\left(\frac{\ln c}{\sqrt{c}}\right),$$

which tends rapidly to 1 as $c \to \infty$.

For example, for $c = 2 \times 10^7$ and base $b = 2$ we find

$$\pi(c) = 1{,}270{,}607, \qquad P_2(c) = 1{,}016,$$

so

$$\text{Accuracy}_2(2 \times 10^7) = \frac{1{,}270{,}607}{1{,}270{,}607 + 1{,}016} \approx 0.99920 \quad (99.92\%).$$

Even for base 2, the Fermat test is therefore extremely accurate in practice, and base 7 performs even better on this range.

5.5   Mersenne Primes

Figure 5.3: Marin Mersenne (1588–1648)

We learned about Fermat tests for different bases and saw that they work very well to check selected primes. So if you pick a huge number at random and only one Fermat test passes, it is highly likely that the number is prime, especially for large numbers.

Now, can we find numbers that automatically satisfy a Fermat test for a specific base? The answer is yes.

Mersenne Numbers are of the form $M_n = 2^n - 1$. If n is composite, M_n is certainly composite. If n is prime, then $2^n - 1 \equiv 1 \mod n$, which often suggests M_n might be prime. If both n and M_n are prime, M_n is called a Mersenne prime M_p.

Hence, all numbers $2^p - 1$ pass the Fermat test base 2. As we observed, if a number passes one Fermat test, it is likely prime, especially for large numbers. However, not all $2^p - 1$ are prime.

To filter them further, we can add another Fermat test, say base 3, for a stronger check. Consider

p prime and $c = 2^p - 1$. If $3^{c-1} \bmod c = 1$, then c is highly likely prime. A simple pseudocode:

```
Initialize an empty list called mersenne_candidates

For each prime number p in primes_3_20k:
    c = 2^p - 1
    if 3^(c-1) mod c == 1:
        append p to mersenne_candidates

print mersenne_candidates
```

The output is:

$$[3, 5, 7, 13, 17, 19, 31, 61, 89, 107, 127, 521, 607, 1279, 2203, 2281,$$
$$3217, 4253, 4423, 9689, 9941, 11213, 19937]$$

All of them, without any missing, are Mersenne primes. This indicates that the Fermat test is highly effective and as we already seen it becomes even more powerful for larger numbers in this family.

Discovering new big primes is an interesting field. It is one of the universal metrics that shows human knowledge power.

If you are fascinated by very large prime numbers, this is for you. We expect that the first prime exponent p after $332{,}192{,}810$ for which

$$3^{\,2^p - 2} \equiv 1 \quad (\bmod \ 2^p - 1)$$

occurs will make $2^p - 1$ an extremely strong prime candidate—indeed, it would have more than $100{,}000{,}000$ decimal digits (since $\#\text{digits}(2^p - 1) = \lfloor p \log_{10} 2 \rfloor + 1$).

Computing such remainders for numbers of this size requires powerful processors and highly efficient modular-exponentiation algorithms and we need to discover an elegant relation between numbers.

5.6   Carmichael Numbers

There are infinitely many composite numbers that can pass all the Fermat tests for all $\gcd(m, c) = 1$:

$$m^{c-1} \bmod c = 1.$$

These are known as Carmichael numbers.

The reason we have these numbers is that their period T is a fraction of $c - 1$, i.e., $T = \frac{c-1}{k}$ with $k \geq 2$. From the period column, the column m^T is all ones (for $\gcd(m, c) = 1$), and so are the columns $m^{2T}, m^{3T}, \dots$. Thus, they pass their ones to multiples of the period. Hence, a Carmichael number c has all ones in the column $m^{c-1} \bmod c$ for all m with $\gcd(m, c) = 1$, because $c - 1 = kT$.

So if we test c with some m and find $m^{c-1} \equiv 1 \bmod c$, it might still be prime, a Carmichael number, or just another composite. Testing all $m \leq \sqrt{c}$ can be hard for large c.

If a number c is a Carmichael number, then $T(c) \mid (c - 1)$ but $T(c) \neq (c - 1)$. If a number c is a prime number, then $T(c) = (c - 1)$.

5.7 Prime tests

There exist many different heuristic and deterministic prime tests.

We know a number c is prime if and only if it is not divisible by any primes $\leq \sqrt{c}$.

The number of operations depends on the number of primes below $\sqrt{c}$, approximated as

$$\sqrt{c}/\ln(\sqrt{c}).$$

This is a growing function, meaning we encounter problems when calculating the primality of large numbers.

Let's again look at the table $m^n \mod 7$. Column 0, except row 0, is all ones. Any 1 we see in the table is a replicate of these ones that appear there by a defined period per row. The combined period $T(c)$ can be calculated by the Carmichael function. So there is nothing random; the key is in the period.

For prime numbers the period is

$$T(c) = c - 1.$$

$T(c)$ can be calculated by the Carmichael function. The Carmichael function is straightforward if we factorize c, and if we already have the factors of c we are not really looking for the primality of c anymore! So we need to find another solution. We need to find other ways to understand whether a number is prime or even to calculate the Carmichael function. Let's again go deep into the exponential remainder table. The period is the key.

$m^n \mod 7$	m^0	m^1	m^2	m^3	m^4	m^5	m^6	m^7	m^8	m^9	m^{10}	m^{11}	m^{12}	m^{13}	m^{14}	m^{15}
0	1	0	0	0	0	0	0	0	0	0	0	0	0	0	0	0
1	1	1	1	1	1	1	1	1	1	1	1	1	1	1	1	1
2	1	2	4	1	2	4	1	2	4	1	2	4	1	2	4	1
3	1	3	2	6	4	5	1	3	2	6	4	5	1	3	2	6
4	1	4	2	1	4	2	1	4	2	1	4	2	1	4	2	1
5	1	5	4	6	2	3	1	5	4	6	2	3	1	5	4	6
6	1	6	1	6	1	6	1	6	1	6	1	6	1	6	1	6

Fermat test. If for a base b coprime to c we have

$$b^{c-1} \equiv 1 \quad (\mod c),$$

then we say that the number c passes the Fermat test to base b. This is a strong sign that c could be prime, but it is incomplete.

The question is: how can we make the Fermat test complete?

A number c is prime if, for all primes $\leq \sqrt{c}$ used as bases p,

$$p^{c-1} \equiv 1 \quad (\mod c),$$

then c has no prime factors $\leq \sqrt{c}$.

Example $c = 7$. Primes $\leq \sqrt{7}$: $\{2\}$.

$$2^6 \mod 7 = 1.$$

So 7 is prime.

The drawback is that the number of operations is not improved.

Let's look from the row side.

If there exists a base b such that the set $\{b^i \mod c; i = 0 \text{ to } (c-1)\}$ has c distinct elements, then c is prime.

So if there exists a base b such that for all $n : 1$ to $(c-2)$ we have $b^n \mod c \neq 1$, then c is prime. Let's prove that 7 is prime. Let's choose $b = 3$.

$b^n \mod 7$:

$$
\begin{aligned}
3^0 \mod 7 &= 1, \\
3^1 \mod 7 &= 3 * (1) \mod 7 = 3, \\
3^2 \mod 7 &= 3 * (3) \mod 7 = 2, \\
3^3 \mod 7 &= 3 * (2) \mod 7 = 6, \\
3^4 \mod 7 &= 3 * (6) \mod 7 = 4, \\
3^5 \mod 7 &= 3 * (4) \mod 7 = 5.
\end{aligned}
$$

$\{1, 3, 2, 6, 4, 5\}$

So the period is 6 and $c = 7$ is a prime.

The drawback is that the number of operations is $O(n)$.

The important question is whether the period is genuinely shaped on $(c-1)$ or not. If yes, the number c is prime; if no, the number is composite.

Let's assume a number c passes the Fermat test base b.

Let's check the genuineness of the period.

Let $F(c-1)$ be the set of prime factors of $(c-1)$. We define the set of top divisors as

$$
D(c-1) = \{(c-1)/f_i \text{ for } f_i \in F(c-1)\}.
$$

Let's assume the period $T(c)$ has formed in column $(c-1)$. This is equivalent to saying that the period does not occur in any of the columns corresponding to the top divisors of $(c-1)$.

Equivalently: if the period does not appear in any of the top divisors $D(c-1)$, then the number c is prime. We may achieve the result with just several checks.

The drawback is that we need to find the prime factors of $(c-1)$, which is pleasant for numbers $m^n + 1$ when the factors of m are known.

This idea is very close in spirit to the classical primality criterion of Édouard Lucas, outstanding mathematician who developed one of the first systematic tests based on the behaviour of powers modulo c.

Figure 5.4: Édouard Lucas (1842–1891), pioneer of sequence–based primality tests

Examples

c=11.

$$c - 1 = 10 = 2 \cdot 5,$$

$$F(10) = \{2, 5\},$$

$$D(10) = \left\{ \tfrac{10}{2}, \tfrac{10}{5} \right\} = \{5, 2\}.$$

Using base 2:

$$2^{10} \equiv 1 \pmod{11},$$

$$2^5 \equiv 10 \not\equiv 1 \pmod{11},$$

$$2^2 \equiv 4 \not\equiv 1 \pmod{11}.$$

Thus the period cannot sit inside exponents 5 or 2, and by the criterion we conclude that 11 is prime.

c=907.

$$c - 1 = 906 = 2 \cdot 3 \cdot 151,$$

$$F(906) = \{2, 3, 151\},$$

$$D(906) = \left\{ \tfrac{906}{2}, \tfrac{906}{3}, \tfrac{906}{151} \right\} = \{453, 302, 6\}.$$

Using base 2:

$$2^{906} \equiv 1 \pmod{907},$$

$$2^{6} \equiv 64 \not\equiv 1 \pmod{907},$$

$$2^{302} \equiv 384 \not\equiv 1 \pmod{907},$$

$$2^{453} \equiv 906 \not\equiv 1 \pmod{907}.$$

So the period does not appear in any of the top exponents, and therefore 907 is prime.

c=9587.

$$c - 1 = 9586 = 2 \cdot 4793,$$

$$F(9586) = \{2, 4793\},$$

$$D(9586) = \left\{ \tfrac{9586}{2}, \tfrac{9586}{4793} \right\} = \{4793, 2\}.$$

Using base 2:

$$2^{9586} \equiv 1 \pmod{9587},$$

$$2^{4793} \equiv 9586 \not\equiv 1 \pmod{9587},$$

$$2^{2} \equiv 4 \not\equiv 1 \pmod{9587}.$$

Thus 9587 is prime.

c=1361.

$$c - 1 = 1360 = 2^4 \cdot 5 \cdot 17,$$

$$F(1360) = \{2, 5, 17\},$$

$$D(1360) = \left\{ \tfrac{1360}{2}, \tfrac{1360}{5}, \tfrac{1360}{17} \right\} = \{680, 272, 80\}.$$

Using base 2:

$$2^{1360} \equiv 1 \pmod{1361},$$

$$2^{80} \equiv 316 \not\equiv 1 \pmod{1361},$$

$$2^{272} \equiv 211 \not\equiv 1 \pmod{1361}.$$

Using base 3:

$$3^{680} \equiv 1360 \not\equiv 1 \pmod{1361}.$$

So for each $e \in D(1360)$ we have a base b_e that rules out a shorter period, and 1361 is prime.

c=341.

$$c - 1 = 340 = 2^2 \cdot 5 \cdot 17,$$

$$F(340) = \{2, 5, 17\},$$

$$D(340) = \left\{ \tfrac{340}{2}, \tfrac{340}{5}, \tfrac{340}{17} \right\} = \{170, 68, 20\}.$$

Using base 3:

$$3^{340} \equiv 56 \not\equiv 1 \pmod{341},$$

$$\gcd(3, 341) = 1.$$

So 341 fails even the basic Fermat test to base 3, and is composite and not a Carmichael number.

c=645.

$$c - 1 = 644 = 2^2 \cdot 7 \cdot 23,$$

$$F(644) = \{2, 7, 23\},$$

$$D(644) = \left\{ \tfrac{644}{2}, \tfrac{644}{7}, \tfrac{644}{23} \right\} = \{322, 92, 28\}.$$

Using base 3:

$$3^{644} \equiv 36 \not\equiv 1 \quad (\bmod\ 645),$$

$$3^{322} \equiv 264 \not\equiv 36 \quad (\bmod\ 645),$$

$$3^{92} \equiv 111 \not\equiv 36 \quad (\bmod\ 645),$$

$$3^{28} \equiv 6 \not\equiv 36 \quad (\bmod\ 645).$$

Therefore 645 is composite; and since for every $e \in \{322, 92, 28\} \subset D(644)$ we have $3^e \not\equiv 3^{644}$ (mod 645), it is not a Carmichael number. Also, as $\gcd(3, 645) = 1$ and $3^{644} \not\equiv 1$ (mod 645), we see that $c = 645$ is not a Carmichael number.

c=1387.

$$c - 1 = 1386 = 2 \cdot 3^2 \cdot 7 \cdot 11,$$

$$F(1386) = \{2, 3, 7, 11\},$$

$$D(1386) = \left\{ \tfrac{1386}{2}, \tfrac{1386}{3}, \tfrac{1386}{7}, \tfrac{1386}{11} \right\} = \{693, 462, 198, 126\}.$$

Using base 2:

$$2^{1386} \equiv 1 \quad (\bmod\ 1387),$$

so c passes the Fermat test base 2. If the exponential period $T(c)$ were exactly formed on $(c - 1)$ then c would be prime, so we investigate the top exponents:

$$2^{693} \equiv 512 \not\equiv 1 \quad (\bmod\ 1387),$$

$$2^{462} \equiv 1322 \not\equiv 1 \quad (\bmod\ 1387),$$

$$2^{198} \equiv 1 \quad (\bmod\ 1387),$$

$$2^{126} \equiv 1 \quad (\bmod\ 1387).$$

So there is a shorter exponent dividing both 198 and 126. Switching to base 3:

$$3^{1386} \equiv 875 \not\equiv 1 \quad (\bmod\ 1387),$$

$$3^{126} \equiv 875 \quad (\bmod\ 1387),$$

$$3^{198} \equiv 875 \quad (\bmod\ 1387).$$

Thus 1387 is composite and, as $\gcd(3, 1387) = 1$, not a Carmichael number. Further checks with base 5 also confirm non-Carmichael:

$$5^{1386} \equiv 1122 \quad (\bmod\ 1387),$$

$$5^{126} \equiv 1141 \not\equiv 1122 \quad (\bmod\ 1387),$$

$$5^{198} \equiv 1141 \not\equiv 1122 \quad (\bmod\ 1387).$$

Calculate the Carmichael function by factorization of $(c - 1)$.

c=561.

$$c - 1 = 560 = 2^4 \cdot 5 \cdot 7,$$

$$F(560) = \{2, 5, 7\},$$

$$D(560) = \left\{ \tfrac{560}{2}, \tfrac{560}{5}, \tfrac{560}{7} \right\} = \{280, 112, 80\}.$$

Using base 2:

$$2^{560} \equiv 1 \quad (\mathrm{mod}\ 561),$$

$$2^{80} \equiv 1 \quad (\mathrm{mod}\ 561),$$

$$2^{112} \equiv 103 \not\equiv 1 \quad (\mathrm{mod}\ 561),$$

$$2^{280} \equiv 1 \quad (\mathrm{mod}\ 561).$$

Using base 3:

$$3^{560} \equiv 375 \not\equiv 1 \quad (\mathrm{mod}\ 561),$$

$$3^{80} \equiv 375 \quad (\mathrm{mod}\ 561),$$

$$3^{280} \equiv 441 \not\equiv 375 \quad (\mathrm{mod}\ 561),$$

$$3^{560} \equiv 3^{80} \equiv 375 \quad (\mathrm{mod}\ 561).$$

Using bases 11 and 17:

$$11^{560} \equiv 11^{80} \equiv 154 \quad (\mathrm{mod}\ 561),$$

$$17^{560} \equiv 17^{80} \equiv 34 \quad (\mathrm{mod}\ 561).$$

One checks similarly that

$$p^{80} \equiv 1 \quad (\mathrm{mod}\ 561) \quad \text{for all remaining primes } p.$$

Therefore 561 is a Carmichael number. It also says that the period $T(561)$ occurs in column 80, in other words $T \mid 80$. To achieve the exact value of the period $T(561)$ we need to recursively continue with the top divisors of 80, $D(80) = \{40, 16\}$. If we show the period is not shaped in $\{16, 40\}$, we prove that $T = 80$.

$$3^{560} \equiv 375 \not\equiv 1 \quad (\mathrm{mod}\ 561),$$

$$3^{80} \equiv 375 \quad (\mathrm{mod}\ 561),$$

$$3^{40} \equiv 441 \not\equiv 375 \quad (\mathrm{mod}\ 561),$$

$$3^{16} \equiv 69 \not\equiv 375 \quad (\mathrm{mod}\ 561).$$

So the period $T(561)$ is 80. Based on the Carmichael function we reviewed in the previous chapter,

$$T(561) = \lambda(561) = 80.$$

Challenge. Decide, using this criterion, whether

$$c = 2^{16} + 1$$

is prime or composite.

Chapter 6

Prime Configuration Framework

"Seventy and two long years, day and night,
I gathered wisdom with all mortal might.
And now I know for certain, after this chase:
None knows life's secret — not a single trace."

— Khayyam

Figure 6.1: Ibn al-Haytham (Alhazen) (c. 965–1040)

6.1 Preface

In Chapter One we introduced the *Atomic Model of Numbers*. Every number is modeled like an atom, with particles named *primons* orbiting the nucleus. In this chapter we take the positions of those primons and encode them in a new arithmetic object, the *prime configuration*, denoted $PC()$.

Given the prime configuration of a number n, we not only learn the prime status of n itself, but also gain information about the prime status of its neighbors, at least up to about n^2. There is a kind of magic here: we can tell whether a large number has any prime factor less than n. The way we map the primons' positions into a configuration is therefore crucial.

We can map them with the usual modulus function: for a primon p we have positions in $[0, 1, \ldots, p - 1]$. A much better mapping is obtained using the *symmetric* modular function. In this short chapter, we build the mathematics behind the *Prime Configuration Framework*, denoted $PC()$.

We dedicate this chapter to Edsger W. Dijkstra and Harvey Dubner—two pioneers of computational thinking, one through rigorous correctness and algorithmic clarity, the other through fearless prime hunting and practical high-performance computing—whose spirit of precision and exploration powers modern discovery in number theory.

Ibn al-Haytham (Alhazen) (c. 965–1040)

Ibn al-Haytham was a groundbreaking mathematician, physicist, and philosopher whose work transformed optics and reshaped the methodology of scientific inquiry. In his monumental *Book of Optics*, he rejected purely speculative reasoning and established a systematic, experimental approach grounded in observation, measurement, and geometric analysis. He treated light as a physical phenomenon governed by precise laws and used mathematical reasoning to explain vision, reflection, and refraction. For this reason, he is often regarded as one of the earliest architects of the scientific method.

In mathematics, Ibn al-Haytham made important contributions to number theory and analysis. He investigated problems involving sums of powers of integers, deriving formulas for

$$1^2 + 2^2 + \cdots + n^2 \quad \text{and} \quad 1^3 + 2^3 + \cdots + n^3,$$

work that later influenced the development of integral calculus. He also studied properties of perfect numbers and divisibility, engaging with questions closely connected to prime numbers. In particular, his investigations into even perfect numbers relied on the structure

$$2^{p-1}(2^p - 1),$$

where $2^p - 1$ is prime—an expression that links perfect numbers directly to what we now call Mersenne primes.

His mathematical style was geometric, structural, and rigorously logical: complex problems were decomposed into simpler components and reconstructed through careful deduction. This intellectual discipline parallels the study of prime numbers today. Just as he analyzed light through precise mathematical frameworks, modern number theory seeks to illuminate hidden structure beneath apparent irregularity. His legacy lies not only in specific results, but in the conviction that clarity, structure, and proof are the foundation of discovery.

Edsger W. Dijkstra (1930–2002)

Edsger W. Dijkstra was a Dutch pioneer of computer science whose work helped turn programming from a craft into a mathematical discipline. He championed structured programming, program correctness, and clear notation, insisting that algorithms should be proved right, not just tested. Many of the core ideas we use today—such as shortest–path algorithms on graphs, disciplined use of recursion, and reasoning about state—grow out of his way of thinking about computation as a precise, logical process. This same spirit applies to prime numbers and number theory: when we design frameworks for primes or new ways to classify numbers, we are following Dijkstra's ideal that complex systems (whether programs, algorithms, or primes themselves) should be built from simple, well–understood components, connected by clean and rigorous reasoning.

Figure 6.2: Edsger W. Dijkstra (1930–2002)

Harvey Dubner (1928–2019)

Harvey Dubner (1928–2019) was an American electrical engineer and legendary "prime hunter" who proved how far curiosity and clever computing can take a single determined mind. Working largely outside academia, he helped pioneer practical, low-cost high-performance computation for exploring big numbers (including the well-known "Dubner Cruncher," built with his son Robert), and he became famous for uncovering and popularizing

Figure 6.3: Harvey A. Dubner (1928–2019)

6.2 Prime Configuration Framework (Overview)

The Prime Configuration Framework $PC()$ assigns to each number a finite sequence that acts as a *signature*. By inspecting the prime configuration of a number we can decide whether it is prime or composite. At the same time, we obtain information about the status of its neighbors, and we can see whether larger numbers have any prime factor that appears in the configuration of n.

6.2.1 Domain Primes and Symmetric Remainder

For a domain parameter $n \geq 2$, let

$$P(n) = [\,2, 3, \ldots, p_s\,]$$

be the list of all primes $\leq n$ (so p_s is the largest prime $\leq n$).

For a modulus $m \geq 2$, define the *symmetric remainder* $a \bmod {}_{\text{sym}} m$ to be the unique integer r such that

$$-\frac{m}{2} < r \leq \frac{m}{2} \qquad \text{and} \qquad a \equiv r \pmod{m}.$$

(For odd m this picks $r \in \{-(m-1)/2, \ldots, (m-1)/2\}$; for $m = 2$ it picks $r \in \{0, 1\}$.)

We may use the usual remainder, the symmetric remainder, or any other consistent remainder system (for example a negative remainder system). In this book we mostly prefer the symmetric remainder.

6.2.2 Definition: Prime Configuration of a Number

Definition. Given a domain n and any integer x, the *Prime Configuration of x relative to n* is

$$PC_n(x) = \big[\, x \bmod p \,\big]_{p \in P(n)}.$$

Equivalently, using the symmetric remainder system,

$$PC_n(x) = \big[\, x \bmod {}_{\text{sym}}p \,\big]_{p \in P(n)}.$$

When $x = n$, we write

$$PC_n := PC_n(n),$$

and, when the domain is clear from context or chosen to be $n = x$, we also write simply $PC(x)$.

6.2.3 Arithmetic in Configuration Space

Arithmetic in the configuration space. Let $k, r \in \mathbb{Z}$, and fix a domain n. Then

$$PC_n(kn + r) = \big(k \cdot PC_n + r\big) \bmod {}_{\text{sym}}P(n),$$

where the operations are performed component-wise over the prime list $P(n)$. Addition, subtraction, and multiplication by scalars all act component-wise, so multiple streams can be advanced independently in parallel and later merged by aligning domains.

6.2.4 Classification Window

Proposition 1 (Exact classification up to the next prime square). Let p_s be the largest prime $\leq n$ and p_{s+1} the next prime. If $m < p_{s+1}^2$ and $PC_n(m)$ contains no zero entry, then m is prime. Conversely, if $PC_n(m)$ has a zero at the coordinate for some $p \in P(n)$, then $p \mid m$ and m is composite.

Every composite $m < p_{s+1}^2$ has a prime factor $\leq p_s$; such a factor appears as a zero in $PC_n(m)$. The converse is immediate from the definition.

6.3 Worked Examples

All configurations below are computed using the symmetric-remainder convention above. For brevity we write $PC(x)$ for $PC_x(x)$, i.e. the prime configuration with domain $n = x$.

$$PC(5) = \big[5 \bmod {}_{\text{sym}}2, \; 5 \bmod {}_{\text{sym}}3, \; 5 \bmod {}_{\text{sym}}5\big] = [1, -1, 0],$$

$$PC(20) = [0, -1, 0, -1, -2, -6, 3, 1] \qquad (\bmod\, 2, 3, 5, 7, 11, 13, 17, 19),$$

$$PC(31) = [1, 1, 1, 3, -2, 5, -3, -7, 8, 2, 0] \qquad (\bmod\, 2, 3, \ldots, 31),$$

$$PC(32) = [0, -1, 2, -3, -1, 6, -2, -6, 9, 3, 1],$$

$$PC(35) = [1, -1, 0, 0, 2, -4, 1, -3, -11, 6, 4].$$

In particular, for a prime x with domain $n = x$, the configuration $PC(x)$ has exactly one zero, occurring in the final position (the coordinate for $p_s = x$).

6.4 Prime Configuration Over Time

Let $PC(t, n) := PC_n(t)$. If $t = kn + r$ with $k, r \in \mathbb{Z}$, then

$$PC(t, n) = (k \cdot PC_n + r) \bmod {}_{\text{sym}} P(n).$$

Thus, once PC_n is known, every residue vector $PC(t, n)$ with the same domain is obtained by a cheap component-wise update. In particular, for all $t < p_{s+1}^2$, if $PC(t, n)$ has no zero entry then t is prime.

Challenge

Compute $PC(10)$; then use it to decide whether 13 and 97 are prime.

$$P(10) = [2, 3, 5, 7], \qquad PC(10) = PC_{10}(10) = [0, 1, 0, 3].$$

Domain $n = 10$ guarantees exact classification up to $p_{s+1}^2 = 11^2 = 121$.

$$\begin{aligned} PC_{10}(13) &= (PC(10) + 3) \bmod {}_{\text{sym}} P(10) = [3, 4, 3, 6] \bmod {}_{\text{sym}} [2, 3, 5, 7] \\ &= [1, 1, -2, -1], \end{aligned}$$

which has no zero; hence 13 is prime.

$$\begin{aligned} PC_{10}(97) &= (9 \cdot PC(10) + 7) \bmod {}_{\text{sym}} P(10) \\ &= (9 \cdot [0, 1, 0, 3] + 7) \bmod {}_{\text{sym}} [2, 3, 5, 7] \\ &= [7, 16, 7, 34] \bmod {}_{\text{sym}} [2, 3, 5, 7] = [1, 1, 2, -1], \end{aligned}$$

again with no zero; hence 97 is prime.

$$P(35) = [2, 3, 5, 7, 11, 13, 17, 19, 23, 29, 31], \qquad PC(35) = [1, -1, 0, 0, 2, -4, 1, -3, -11, 6, 4].$$

Here $p_{s+1} = 37$, so the window extends at least to $35^2 = 1225$ and, in fact, to $37^2 = 1369$.

$$\begin{aligned} PC_{35}(91) &= (2 \cdot PC(35) + 21) \bmod {}_{\text{sym}} P(35) = [1, 1, 1, 0, 3, 0, 6, -4, -1, 4, -2], \\ &\text{zeros at } 7, 13 \Rightarrow 91 = 7 \cdot 13; \end{aligned}$$

$$\begin{aligned} PC_{35}(1111) &= (31 \cdot PC(35) + 26) \bmod {}_{\text{sym}} P(35) = [1, 1, 1, -2, 0, 6, 6, 9, 7, 9, -5], \\ &\text{zero at } 11 \Rightarrow 1111 \text{ is composite}; \end{aligned}$$

$$\begin{aligned} PC_{35}(1001) &= (28 \cdot PC(35) + 21) \bmod {}_{\text{sym}} P(35) = [1, -1, 1, 0, 0, 0, -2, -6, -11, -14, 9], \\ &\text{zeros at } 7, 11, 13 \Rightarrow 1001 = 7 \cdot 11 \cdot 13; \end{aligned}$$

$$\begin{aligned} PC_{35}(7203) &= (205 \cdot PC(35) + 28) \bmod {}_{\text{sym}} P(35) = [1, 0, -2, 0, -2, 1, -5, 2, 4, 11, 11], \\ &\text{zeros at } 3, 7 \Rightarrow 7203 \text{ is composite}; \end{aligned}$$

$$\begin{aligned} PC_{35}(1009) &= (28 \cdot PC(35) + 29) \bmod {}_{\text{sym}} P(35) = [1, 1, -1, 1, -3, -5, 6, 2, -3, -6, -14], \\ &\text{no zero} \Rightarrow 1009 \text{ is prime}. \end{aligned}$$

The Prime Factorization Engine in *NumbersDiscoveries.com* is based on the Prime Configuration Framework .

6.5   Next Prime Number

What is the next prime number after $n = 33$?

Let p_s be the largest prime $\leq n$. We want to find p_{s+1}. We need the configuration using primes $\leq \sqrt{n}$ and then extend it until we pass the next prime. Once a candidate $m = n + q$ is found with $m < 7^2 = 49$, the configuration with primes ≤ 5 is sufficient to certify primality.

We compute

$$PC_5(33) = 33 \bmod [2, 3, 5] = [1, 0, 3].$$

The "next–zero distances" for each wave are

$$[2, 3, 5] - [1, 0, 3] = [1, 3, 2],$$

meaning the next crossings after 33 occur at offsets

$$\text{for } 2 : \; 1, 3, 5, 7, 9, 11, \ldots$$
$$\text{for } 3 : \; 3, 6, 9, 12, \ldots$$
$$\text{for } 5 : \; 2, 7, 12, 17, \ldots$$

Let $Q = \{1, 2, 3, 4, 5, 6, 7, 8, 9, 10, 11, 12, \ldots\}$ be all positive offsets. Removing the offsets hit by any wave gives the prime-gap candidates.

First candidate. The union of hits contains $\{1, 2, 3, 5, 6, 7, 9, 11, 12, \ldots\}$, so the smallest offset not hit is $q = 4$. Therefore $m = 33 + 4 = 37$. Since $37 < 49 = 7^2$, 37 is prime.

Second candidate. Continue the same lists (they already contain all future hits). The next offset not in $\{1, 2, 3, 5, 6, 7, 9, 11, 12, \ldots\}$ is $q = 8$. Thus $m = 33 + 8 = 41$; again $41 < 49$, so 41 is prime.

Third candidate. The next missing offset is $q = 10$. Thus $m = 33 + 10 = 43$; since $43 < 49$, we conclude 43 is also prime.

In summary, starting from $n = 33$ with $PC_5(33) = [1, 0, 3]$, the successive prime offsets are $q = 4, 8, 10$, giving the primes $37, 41, 43$.

6.6   Prime Factorization Engine

If we know PC_n (the prime configuration at domain n with prime list $P(n) = [2, 3, \ldots, p_s]$), we can certify primality and detect composite numbers correctly for all $m < p_{s+1}^2$. In particular, $PC_n(m)$ immediately reveals whether m has any prime factor $\leq p_s$. This is the core power of the Prime Configuration Framework $PC()$.

To navigate large ranges efficiently, we place *checkpoints* on the number line—fixed lights in the dark—at which we precompute and store the configuration:

$$L_2 = \left[2^2, 2^4, 2^8, 2^{16}, 2^{32}, 2^{64}, 2^{128}, \ldots \right],$$

$$\text{or in base 10} \quad L_{10} = \left[10, 10^2, 10^4, 10^8, 10^{16}, 10^{32}, 10^{64}, 10^{128}, \ldots \right].$$

At each checkpoint $\ell \in L_2$ (or L_{10}) we record $PC(\ell)$. Then any nearby t can be expressed as $t = k\ell + r$ with integers k, r, and updated by

$$PC(t, \ell) \;=\; \left(k \cdot PC(\ell) + r \right) \bmod {}_{\text{sym}} P(\ell).$$

This is just three vector operations (scale, add, symmetric–mod), and it parallelizes naturally. The resulting $PC(t, \ell)$ shows all prime factors of t that lie in $P(\ell)$, and—whenever $t < p_{s(\ell)+1}^2$—it is sufficient to determine the prime/composite status of t.

6.7 Generating Primes Using Zero

Zero or not zero—that is the question!

6.7.1 Initial Table Setup

We construct a table where:

- The columns represent natural numbers starting from 2; column n corresponds to the number n.

- The leftmost column (index column) lists the primes discovered so far.

6.7.2 Fundamental Rule

Every column must contain at least one zero. If a column has no zero:

1. append a zero to the column,

2. declare the corresponding number to be prime,

3. and add it to the prime index.

   To fill out subsequent columns:

1. Compute the next column by adding 1 to the previous column.

2. If any cell value equals the corresponding prime in the index column, replace it with 0.

   These rules are enough to generate all prime numbers sequentially.
   Let us generate primes using this method.

primes	2	3	4	5	6	7	8	9	10	11	12	13	14	15	16	17

Step 1. Start with the number 2. The second column has no zeros, so we add a zero and declare 2 to be prime. We then add 2 to the prime index.

primes	2	3	4	5	6	7	8	9	10	11	12	13	14	15	16	17
2	0															

Step 2. Move to the next number, 3. The third column is obtained by adding 1 to the previous column (which corresponds to 2).

primes	2	3	4	5	6	7	8	9	10	11	12	13	14	15	16	17
2	0	1														

Since there is still no zero in this column, we add a zero and declare 3 to be prime, adding it to the prime index.

primes	2	3	4	5	6	7	8	9	10	11	12	13	14	15	16	17
2	0	1														
3		0														

Step 3. Proceed to the number 4. Add 1 to the previous column (corresponding to 3). If the new value equals a prime in the index column, replace it with 0. In this case, we replace 2 (in the row for prime 2) by 0.

primes	2	3	4	5	6	7	8	9	10	11	12	13	14	15	16	17
2	0	1	0													
3		0	1													

Step 4. Continue with the number 5. Add 1 to the previous column (corresponding to 4).

primes	2	3	4	5	6	7	8	9	10	11	12	13	14	15	16	17
2	0	1	0	1												
3		0	1	2												

Since there is no zero in this new column yet, we append a zero, declare 5 to be prime, and add it to the index.

primes	2	3	4	5	6	7	8	9	10	11	12	13	14	15	16	17
2	0	1	0	1												
3		0	1	2												
5				0												

6.8 Generating Prime Numbers up to 17

Continuing this process, we obtain the following table for numbers up to 17:

primes	2	3	4	5	6	7	8	9	10	11	12	13	14	15	16	17
2	0	1	0	1	0	1	0	1	0	1	0	1	0	1	0	1
3		0	1	2	0	1	2	0	1	2	0	1	2	0	1	2
5				0	1	2	3	4	0	1	2	3	4	0	1	2
7						0	1	2	3	4	5	6	0	1	2	3
11										0	1	2	3	4	5	6
13												0	1	2	3	4
17																0

6.9 Properties

Let us look inside the table and highlight some of the structure hidden there.

6.9.1 Prime Counting Function

The prime-counting function $\pi(n)$ counts the number of primes less than or equal to a given number n.

In this framework, the *height* of each column (the number of rows present in that column) is exactly $\pi(n)$. For example, the column for 17 has rows indexed by

$$2, 3, 5, 7, 11, 13, 17,$$

so its height is 7, and indeed $\pi(17) = 7$.

primes	2	3	4	5	6	7	8	9	10	11	12	13	14	15	16	17
2	0	1	0	1	0	1	0	1	0	1	0	1	0	1	0	1
3		0	1	2	0	1	2	0	1	2	0	1	2	0	1	2
5				0	1	2	3	4	0	1	2	3	4	0	1	2
7						0	1	2	3	4	5	6	0	1	2	3
11										0	1	2	3	4	5	6
13												0	1	2	3	4
17																0

6.9.2 Prime Gaps

Prime gaps appear naturally in this table. Moving along a row corresponds to moving through the numbers, while zeros mark exact multiples of the corresponding prime. The patterns between the zeros encode the gaps between primes.

primes	2	3	4	5	6	7	8	9	10	11	12	13	14	15	16	17
2	0	1	0	1	0	1	0	1	0	1	0	1	0	1	0	1
3		0	1	2	0	1	2	0	1	2	0	1	2	0	1	2
5				0	1	2	3	4	0	1	2	3	4	0	1	2
7						0	1	2	3	4	5	6	0	1	2	3
11										0	1	2	3	4	5	6
13												0	1	2	3	4
17																0

6.9.3 Primes and Zero

A crucial feature is that primes have only *one* zero in their column, and this zero is always the last entry in that column. This gives a simple visual criterion for primality: a column is prime if and only if it contains exactly one zero, at the bottom.

primes	2	3	4	5	6	7	8	9	10	11	12	13	14	15	16	17
2	0	1	0	1	0	1	0	1	0	1	0	1	0	1	0	1
3		0	1	2	0	1	2	0	1	2	0	1	2	0	1	2
5				0	1	2	3	4	0	1	2	3	4	0	1	2
7						0	1	2	3	4	5	6	0	1	2	3
11										0	1	2	3	4	5	6
13												0	1	2	3	4
17																0

6.10   Find the First Prime Number Larger Than 9900

Let us now apply the framework at a larger scale.

First, compute $PC(9900)$; then, using that column, we can compute the nearby columns by adding 1 repeatedly and applying the "reset to zero when hitting a prime" rule.

primes	9900+	1	2	3	4	5	6	7	8	9	10	11	12	13	14	15	16	17	18	19	20
2	0	1	0	1	0	1	0	1	0	1	0	1	0	1	0	1	0	1	0	1	0
3	0	1		0		2		1		0		2		1		0		2		1	
5	0	1				0		2				1		3				2		4	
7	2	3						2				6		1				5		0	
11	0	1						7				0		2				6			
13	7	8						1						7				11			
17	6	7						13						2				6			
19	1	2						8						14				18			
23	10	11						17						0				4			
29	11	12						18										28			
31	11	12						18										28			
37	21	22						28										1			
41	19	20						26										36			
43	10	11						17										27			
47	30	31						37										0			
53	42	43						49													
59	47	48						54													
61	18	19						25													
67	51	52						58													
71	31	32						38													
73	45	46						52													
79	25	26						32													
83	23	24						30													
89	21	22						28													
97	6	7						13													

The first column (9900) has multiple zeros, but the first column *after* that whose entries are all nonzero until the last row—where a new zero is appended—is 9901. Thus the first prime greater than 9900 is

$$9901,$$

and continuing the procedure shows that the second is

$$9907.$$

6.11 Find Primes in a Range

Calculate the prime numbers between 70 and 80.

Let the first house be at $n = 70$, and place a house at every integer. House h corresponds to the number $n + h = 70 + h$.

If a house has a prime number sitting in it, then $70 + h$ is composite (that prime is a factor). If the house is empty, then $70 + h$ is prime.

We put each effective prime into a house. If its target house is already occupied, we jump by one full period (its own value) until we find the next empty house.

With these rules, we calculate the primes in $[70, 80]$.

The effective primes for both the minimum and maximum of the bound are

$$[2, 3, 5, 7].$$

The prime configuration of 70 is

$$PC(70) = [0, 1, 0, 0].$$

The locations of the first "prime houses" are determined by

$$[2, 3, 5, 7] - [0, 1, 0, 0] \bmod [2, 3, 5, 7] = [0, 2, 0, 0].$$

So:

- we give house 0 to prime 2,

- we give house 2 to prime 3,

- prime 5 wants house 0, but it is occupied by prime 2, so it jumps to the next house in its period, house 5,

- prime 7 behaves the same way and moves to house 7.

Initial state.

h	0	1	2	3	4	5	6	7
prime in h	2		3			5		7

House $h = 0$

House 0 is not prime: it has factor 2. So $70 + 0 = 70$ is composite.

Now move 2 to its next house. From $h = 0$ the period of 2 is

$$0 \to 2 \to 4 \to 6 \to \ldots$$

House 2 is occupied by 3, so 2 jumps again to house 4, which is empty.

State after processing $h = 0$:

h	1	2	3	4	5	6	7
prime in h		3		2	5		7

House $h = 1$

House 1 is empty, so it "shapes a prime". That is,

$$70 + 1 = 71$$

is prime.

Next stage (drop house 1, keep the others):

h	2	3	4	5	6	7
prime in h	3		2	5		7

House $h = 2$

House 2 is not prime: it has factor 3. So

$$70 + 2 = 72$$

is composite.

We move prime 3. From $h = 2$ the period of 3 is

$$2 \to 5 \to 8 \to 11 \to \dots$$

House 5 is occupied by 5, so the first empty house is $h = 8$. Prime 3 moves to house 8.

State after processing $h = 2$:

h	3	4	5	6	7	8
prime in h		2	5		7	3

House $h = 3$

House 3 is empty, so it "shapes a prime". That is,

$$70 + 3 = 73$$

is prime.

State before checking $h = 4$ (same houses, just move the pointer):

h	4	5	6	7	8
prime in h	2	5		7	3

House $h = 4$

House 4 is not prime: it has factor 2. So

$$70 + 4 = 74$$

is composite.

Prime 2 jumps forward by period 2. From $h = 4$:

$$4 \to 6 \to 8 \to 10 \to \dots$$

House 6 is empty, so 2 moves to house 6.

State after processing $h = 4$:

h	5	6	7	8
prime in h	5	2	7	3

House $h = 5$

House 5 is not prime: it has factor 5. So

$$70 + 5 = 75$$

is composite.

Prime 5 jumps forward by period 5. From $h = 5$:

$$5 \to 10 \to 15 \to \dots$$

The first house we use in this range is $h = 10$, which is empty, so 5 moves to house 10.

State after processing $h = 5$:

h	6	7	8	9	10
prime in h	2	7	3		5

House $h = 6$

House 6 is not prime: it has factor 2. So

$$70 + 6 = 76$$

is composite.

Prime 2 jumps forward by period 2. From $h = 6$:

$$6 \to 8 \to 10 \to 12 \to \dots$$

House 8 has prime 3; house 10 has prime 5; the first empty house is $h = 12$. So 2 moves to house 12.

State after processing $h = 6$:

h	7	8	9	10	11	12
prime in h	7	3		5		2

House $h = 7$

House 7 is not prime: it has factor 7. So

$$70 + 7 = 77$$

is composite.

Prime 7 jumps forward by period 7. From $h = 7$:

$$7 \to 14 \to 21 \to \dots$$

The first house we use in this range is $h = 14$, which is empty, so 7 moves to house 14.

State after processing $h = 7$:

h	8	9	10	11	12	13	14
prime in h	3		5		2		7

House $h = 8$

House 8 is not prime: it has factor 3. So

$$70 + 8 = 78$$

is composite.

Prime 3 jumps forward by period 3. From $h = 8$:

$$8 \rightarrow 11 \rightarrow 14 \rightarrow 17 \rightarrow \ldots$$

House 11 is empty, so 3 moves to house 11.

State after processing $h = 8$:

h	9	10	11	12	13	14
prime in h		5	3	2		7

House $h = 9$

House 9 is empty, so it shows a prime. That is,

$$70 + 9 = 79$$

is prime.

State before checking $h = 10$:

h	10	11	12	13	14
prime in h	5	3	2		7

House $h = 10$

House 10 has prime 5, so

$$70 + 10 = 80$$

is not prime. We are at the end of the bound, so we stop here.

Result

From the process above, we have:

h	$70 + h$	status
0	70	composite (factor 2)
1	71	**prime**
2	72	composite (factor 3)
3	73	**prime**
4	74	composite (factor 2)
5	75	composite (factor 5)
6	76	composite (factor 2)
7	77	composite (factor 7)
8	78	composite (factor 3)
9	79	**prime**
10	80	composite (factor 5)

Chapter 7

Number–Wave Framework

"If you want to find the secrets of the universe, think in terms of energy, frequency, and vibration."
— Tesla

Figure 7.1: Joseph Fourier (1768–1830).

7.1 Preface

7.1.1 Introduction — All Is About Waves

Everything is about waves. More precisely: everything is a wave, or the effect (the shadow) of a wave.

In this chapter, we establish a powerful framework that connects the real world to the discrete world. Then, standing on the Number–Wave Framework, we begin to formulate prime numbers and prime patterns.

We define operations over waves. We define the *spectrum* of waves as a set of waves. Once we have a spectrum, we can formulate prime patterns as structured shadow phenomena.

We also discuss the challenges, look from different angles, and build intuition. In this book, the solution is not the main goal. The main goal is the path: how we define objects, how we operate on them, and what we learn from the structure.

The Number–Wave Framework opens a gate to the world of prime numbers.

We dedicate this chapter to Joseph Fourier and Adrien-Marie Legendre—two masters of extraction: Fourier for revealing hidden structure by translating complexity into frequencies, and Legendre for measuring the pulse of the primes through early analytic approximations that still guide how we model prime density today.

Joseph Fourier: Turning Complexity into Frequencies

Fourier's greatness is that he taught science a new instinct: when a phenomenon looks tangled in its original form, translate it into waves, and the structure will reveal itself. His ideas made it natural to decompose a complicated "signal" into simple oscillations, and that single shift of viewpoint became a foundation for modern physics, engineering, and computation. In the Number–Wave Framework, this is exactly the spirit we adopt: the integer line is our signal, number-generated waves are our harmonics, and hidden arithmetic structure becomes visible when we stop staring at raw integers and begin listening to their frequencies.

Adrien-Marie Legendre: Measuring the Pulse of Primes

Figure 7.2: Adrien-Marie Legendre (1752–1833).

Legendre's greatness lies in his insistence that primes are not merely a list, but a phenomenon with a discoverable global shape. He pushed number theory toward quantitative thinking—counting, estimating, and extracting laws from data—so that the distribution of primes could be studied like a real natural pattern rather than a mystery of scattered exceptions. For this chapter, Legendre represents the "measurement mindset": once primes are viewed through a structured lens, the right definitions turn chaos into a curve, and the curve becomes something we can compare, refine, and ultimately explain inside a wave-based framework.

7.2 Wave of Numbers

7.2.1 Real Wave (continuous object)

For a nonzero generator T and a phase r, define the real wave on an interval $[s, e]$:

$$\boxed{v_T(x, r, s, e) = \sin\!\left(\frac{\pi}{T}(x - r)\right), \qquad x \in [s, e] \subseteq \mathbb{R}, \quad T \neq 0.}$$

Generator and phase.

- T is the **generator**.

- r is the **phase**, which may be

    - a single residue (scalar),

    - a set,

    - or a real range.

Phase reduction (root-level).   In this framework, as the generators and phases defined as integers so all roots must be integers. so the domain of waves are R but the out is a subset of integers. phase are periodic modulo T:

$$w_T(x, r, s, e) \;=\; w_T(x, r \bmod T, s, e).$$

In our context We only consider positive generators.

7.2.2   Wave of a Number

Integers are the roots represented by the unit wave:

$$\boxed{v_1(x, 0) = \sin(\pi x).}$$

Wave of a number.   For a generator $T \neq 0$, define the integer wave by

$$\boxed{w_T(r, s, e) \;:=\; \{\, x \mid v_T(x, r, s, e) = 0 \,\}.}$$

We call $w_T(x, r, s, e)$ the **wave of number** T with phase r on the domain $[s, e]$.

Wave value at an integer.   The value of the wave at an integer point n is $w_T(x, r, n) \in \{0, 1\}$. It equals 1 if and only if $v_T(x, r, n) = 0$, i.e., n is an integer root of the real wave.

7.2.3   Integer and Prime Waves

- **Integer wave:** if $T \in \mathbb{Z}_{>0}$.

- **Prime wave:** if $T = p$ is prime.

In our context, we work with **integer waves**. When we say *waves*, we mean these integer waves (not the continuous real waves), unless we explicitly say otherwise.

7.2.4   Crossing Points and Shadow Points

A wave crosses some integer points and misses others. The integer points that the wave crosses are the **crossing points** of the wave. All other integer points are **shadow points**. (Throughout, points are integers.)

7.3 Wave Counting Function

We now define the counting function of a wave: the number of integer crossings (roots) inside a finite domain.

7.4 Wave Value Function

$$w_T(x, r, n) = 1 \quad \text{if } v_T(x, r, n) = 0,$$

then n is a **Crossing Point** of $w_T(x, r)$.

$$w_T(x, r, n) = 0 \quad \text{if } v_T(x, r, n) \neq 0,$$

then n is a **Shadow Point** of $w_T(x, r)$.

7.4.1 Wave Counting Function

We use $|\cdot|$ notation for the counting function.

Conceptually, a wave on $[s, e]$ is the union of its point-values:

$$w_T(x, r, s, e) \;=\; \bigcup_{n=s}^{e} w_T(x, r, n).$$

Therefore, the crossing count is the sum of point-values:

$$|w_T(x, r, s, e)| \;=\; \sum_{n=s}^{e} w_T(x, r, n).$$

A standard closed form:

$$|w_T(x, r, s, e)| = \left\lfloor \frac{e - r}{T} \right\rfloor - \left\lfloor \frac{(s - 1) - r}{T} \right\rfloor.$$

We can find the first focal point and rewrite the equation as a single-floor expression:

$$|w_T(x, r, s, e)| = \left\lfloor \frac{T + e - \left(s + ((r - s) \bmod T)\right)}{T} \right\rfloor$$

7.4.2 Point Lumination

The **point lumination** of n is defined as the number of waves crossing at n.

7.5 Probability of Waves

For a finite integer domain $[s, e]$, the empirical wave probability is:

$$P(w_T(x, r, s, e)) = \frac{|w_T(x, r, s, e)|}{e - s + 1}.$$

7.5.1   Independence and Dependency of Waves

Independence.   A wave w_T is called **independent** on $[s, e]$ if and only if the probability of the wave equals the uniform baseline:

$$P\big(w_T(x, r, s, e)\big) = \frac{1}{T}.$$

Unknown-phase independence.   A wave with an unknown phase

$$w_T(x, *, s, e),$$

is treated as independent, since a random integer hits exactly one residue class among T residue classes:

$$P\big(w_T(x, *, s, e)\big) = \frac{1}{T}, \qquad |w_T(x, *, s, e)| = \frac{e - s + 1}{T}.$$

Here $*$ means: a wave exists, but the phase is unknown. We also define $*_k$ to mean: there exist k waves with unknown phases. By convention, $*_{-k}$ is equal to $*_{T-k}$.

Dependency factor.   The deviation from independence is measured by the **dependency factor**:

$$d\big(w_T(x, r, s, e)\big) \ := \ \frac{P\big(w_T(x, r, s, e)\big)}{1/T} = \frac{|w_T(x, r, s, e)|}{(e - s + 1)/T}.$$

Full independence corresponds to $d(w) = 1$.

Different types of independence.

- **Independent wave (unknown phase):**

$$d\big(w_T(x, *, s, e)\big) = 1.$$

- **Independent wave (aligned period):**

$$w_T(x, r, s, e) \text{ is independent if } d\big(w_T(x, r, s, e)\big) = 1,$$

    which occurs whenever

$$T \mid (e - s + 1).$$

- **Asymptotically independent wave:**

$$\lim_{e \to \infty} d\big(w_T(x, r, s, e)\big) = 1 \quad \Longleftrightarrow \quad \lim_{e \to \infty} P\big(w_T(x, r, s, e)\big) = \frac{1}{T}.$$

Asymptotically independent waves are not necessarily independent during the journey from the origin toward infinity. If we assume full independence over a finite domain, then we must calculate the resulting error.

7.5.2 Wave Dependency Factor

We measure deviation from the independent baseline by the dependency factor:

$$
d\big(w_p(x,r,s,e)\big) = \frac{\big|w_p(x,r,s,e)\big|}{\big|w_p(x,*,s,e)\big|} = \frac{\big|w_p(x,r,s,e)\big|}{\frac{(e-s+1)}{p}} = \frac{\left\lfloor \frac{e-s+p-\big((r+s)\bmod p\big)}{p} \right\rfloor}{\frac{(e-s+1)}{p}}
$$

Example. On $[1,n]$ with phase 0:

$$
d\big(w_p(x,0,1,n)\big) = \frac{\big|w_p(x,0,1,n)\big|}{\big|w_p(x,*,1,n)\big|} = \frac{\left\lfloor \frac{n}{p} \right\rfloor}{\frac{n}{p}}.
$$

Here $d = 1$ means perfect independence on that domain.

Bound. Since $\lfloor x \rfloor \le x$ and $x - 1 < \lfloor x \rfloor$, we have

$$
\frac{n}{p} - 1 < \left\lfloor \frac{n}{p} \right\rfloor \le \frac{n}{p}.
$$

Dividing by $\frac{n}{p}$ gives

$$
1 - \frac{p}{n} < d\big(w_p(x,0,1,n)\big) \le 1.
$$

Thus, the dependency factor depends explicitly on both the prime p and the cutoff n.

Effective prime waves. At a point n, the **effective prime waves** are the prime waves that actually matter for the prime status of n. They are exactly:

$$
p_i \le \sqrt{n}, \qquad i = 1, \ldots, s.
$$

The prime status of n is dependent on the crossing/shadow status of these effective prime waves.

Asymptotic independence of effective prime waves. For effective primes $p \le \sqrt{n}$, we have the uniform bound

$$
1 - \frac{\sqrt{n}}{n} < d\big(w_p(x,0,1,n)\big) \le 1.
$$

As $n \to \infty$, the lower bound converges to 1, and hence all effective prime waves are *asymptotically independent*.

7.5.3 Cutoff Error (tail error)

The difference between the independent wave and the actual wave is the **cutoff error** of the wave. Cutoff error for waves can be defined as:

$$
\begin{aligned}
E\big(w_p(x,r,s,e)\big) &= \big|w_p(x,*,s,e)\big| - \big|w_p(x,r,s,e)\big| \\
&= \left\lfloor \frac{e - s + p - \big((r+s)\bmod p\big)}{p} \right\rfloor - \frac{(e-s+1)}{p}
\end{aligned}
$$

The relation between the dependency factor and cutoff error is:

$$d\big(w_p(x,r,s,e)\big) = 1 - \frac{E\big(w_p(x,r,s,e)\big)}{\big|w_p(x,*,s,e)\big|}$$

7.5.4 Prime Waves

A wave w_p is prime if and only if it is coprime with all waves that have smaller generators.

7.5.5 Coprime Waves

Two waves w_p and w_q are coprime if and only if, within one combined period, they each cross all phases of the other wave.

The combined period T_c is generally $p \cdot q$ when the generators are coprime. Otherwise, the combined period is $\mathrm{lcm}(p,q)$.

Example: $w_3(x,1)$ and $w_5(x,4)$ are coprime waves.

$$w_3(x,1) : \{1,4,7,10,13\} \bmod 5 = \{1,4,2,0,3\}$$

$$w_5(x,4) : \{4,9,14\} \bmod 3 = \{1,0,2\}$$

They hit all phases of each other within one combined period, so $w_3(x,1)$ and $w_5(x,4)$ are coprime waves. Also, in practice, the phase of a wave does not affect coprimality and primality; the generator controls it.

In this book, we mostly work with prime waves. They are coprime to each other, independent on their combined period, and asymptotically independent.

There is no way to cover a wave on an unbounded domain using a finite number of other coprime waves.

Let us make this clearer in more physical language. We define the integer frequency of a wave w_T as $1/T$. All of these waves have different frequencies, and it is not possible to cover a wave with a specific frequency using only finitely many other coprime frequencies.

Coprime waves have a clear and strong rotating system: within one combined period, they cross all possible phase-positions. There is no way that—even with infinitely many waves—an unchanging (static) phase pattern can cover a coprime wave over the entire domain.

There are only 3 ways to cover a coprime wave over the whole domain:

1) We cover it with the exact same wave (same generator and the same phase).

2) One of the waves gains all phases and crosses every point in the domain.

3) Using infinitely many waves with dynamic phase assignation. If all the phases be static and they could not complete the space, encountering the focal point is inevitable.

7.6 Spectrum

7.6.1 Definition

A **spectrum** is a set of waves:

$$S(x,T,R,S,E,U) = \{\, w_{T_i}(x,r_i,s_i,e_i) \ : \ x \in U \,\}.$$

In general, each spectrum S has five inputs:

- x is the variable. It can be a scalar or a range.

- T is a range, array, or function of wave generators.

- R is a range, array, or function of wave phases.

- S is a range, array, or function of wave start points.

- E is a range, array, or function of wave end points.

- U is the universal set (generally $\mathbb{Z}, \mathbb{N}, \mathbb{N}_0, \mathbb{N}_2$, etc.).

Generally, for each spectrum all of T, R, S, E, U are defined, and the only variable is x. So for convenience, we write $S(x)$ to refer to $S(x, T, R, S, E, U)$. This makes formulations more concise. If x is an integer n, then:

- If n is a crossing point of the spectrum, the value of the spectrum is 1.

- Otherwise, the value of the spectrum is 0.

So:

- $S(n)$ is the value of the spectrum at point n.

- $S([s, e])$ is the set of crossing points in $[s, e]$.

The value function should be the last function in the process. Otherwise, when x gets replaced by a value, the whole spectrum collapses to a single value. To avoid that, we do not replace x with a value. Instead, x keeps the structure of the waves and spectra, and we set the start and end points of the waves.

Convention.

- $S(n)$ = value of spectrum at n

- $S(s, e)$ = set of crossing points in $[s, e]$

- $S(x, s, e)$ = spectrum S restricted to $[s, e]$

- $S(x, n)$ = spectrum S restricted to $[1, n]$

7.6.2 Crossing points and shadow points

Crossing points. The crossing set (roots of the spectrum) is:

$$Cross(S) = \bigcup_{w \in S} w.$$

Shadow function and shadow points. For any subset $A \subseteq U$, define

$$Shadow(A) := U \setminus A.$$

Therefore, the shadow set of a spectrum is

$$Shadow(S) = Shadow(Cross(S)) = U \setminus Cross(S).$$

Shadow is equivalent to set complement.

7.6.3   Spectrum period

If a spectrum S contains finitely many waves with generators $T_1, \ldots, T_s$, then its crossing structure on the integers is periodic. The **combined period** (spectrum period) is the least common multiple:

$$\boxed{T_c = \operatorname{lcm}(T_1, \ldots, T_s).}$$

In particular, if the generators are distinct primes $q_1, \ldots, q_s$, then

$$\boxed{T_c = q_1 q_2 \cdots q_s.}$$

7.6.4   Complete, Saturated, and Full Spectra

Fix a universal set $U \subseteq \mathbb{Z}$ (typically $U = \mathbb{N}_0$).

Saturated spectrum.   A spectrum S is **saturated** if there exists a point N such that every integer $x \geq N$ is crossed by at least one wave in S. Equivalently, the shadow of S is finite. For example, the single wave

$$S = \{\, w_1(x, 0, 10, \infty) \,\}$$

is right-saturated on $\mathbb{N}_0$ (indeed, it crosses all integers from 10 onward).

Complete spectrum.   A spectrum S is **complete** if every point in the universal set U is crossed by at least one wave in S, i.e., it has no shadow:

$$\operatorname{Shadow}(S) = \emptyset.$$

For example, for modulus 3,

$$C = \{\, w_3(x, 0),\ w_3(x, 1),\ w_3(x, 2) \,\}$$

is complete on U, because every integer is congruent to exactly one residue class modulo 3.

Full spectrum.   A **full spectrum** is the spectrum obtained by taking, for each effective generator T, *all phases* $r \in \{0, 1, \ldots, T - 1\}$. Equivalently, it is the complement (shadow) of the empty spectrum on the same universe. In a full spectrum:

- the number of shadow points in one combined period is 0;

- at every integer point, the number of waves crossing is exactly the number of generators (one phase per generator).

For example, using generators $\{2, 3, 5\}$, the full spectrum is

$$\begin{aligned} F = \{\, &w_2(x, 0), w_2(x, 1),\ w_3(x, 0), w_3(x, 1), w_3(x, 2), \\ &w_5(x, 0), w_5(x, 1), w_5(x, 2), w_5(x, 3), w_5(x, 4) \,\}. \end{aligned}$$

7.6.5 Shadow points by deleting phases from a full spectrum

To create shadow points, we delete phases from the full spectrum. Let the generators be distinct primes $q_1, \ldots, q_s$, with combined period

$$T_c = q_1 q_2 \cdots q_s.$$

For each generator q_i, let d_{q_i} be the number of deleted phases (deleted waves) from that generator. Then the number of shadow points in one combined period is

$$\boxed{\; |\,\mathrm{Shadow}(S) \cap [0, T_c)\,| = \prod_{i=1}^{s} d_{q_i}. \;}$$

(Reason: in $[0, T_c)$, each integer corresponds to a unique phase choice for each generator; a point is a shadow point exactly when it lands in a deleted phase for *every* generator.)

In particular, for generators $\{2, 3, 5\}$ we have $T_c = 30$, and

$$|\,\mathrm{Shadow}(S) \cap [0, 30)\,| = d_2\, d_3\, d_5.$$

First shadow point. The first shadow points can appear only if we delete at least one phase from *each* generator. For example, delete phase 0 from each generator (i.e., keep all nonzero phases). Define

$$G = \{\, w_2(x, 1),\ w_3(x, 1), w_3(x, 2),\ w_5(x, 1), w_5(x, 2), w_5(x, 3), w_5(x, 4)\,\}.$$

Here $d_2 = d_3 = d_5 = 1$, so there is exactly one shadow point per combined period. Indeed, the unique shadow residue in $[0, 30)$ is 0, hence

$$\neg G = w_{30}(x, 0).$$

7.6.6 Lumination

The number of waves of a spectrum S that cross a point n is called the *lumination* of the point n in S.

There are different ways to measure or present lumination, which we will discuss later in the matching problems. For example:

$$l_1(n) = \text{the number of waves crossing } n,$$

$$l_2(n) = \frac{\text{the number of waves crossing } n}{\text{the total number of effective waves}},$$

$$l_3(n) = \sum_{\substack{\text{crossed effective} \\ \text{waves at } n}} 1 - \sum_{\substack{\text{non-crossed effective} \\ \text{waves at } n}} \frac{1}{p_i},$$

where the p_i are the generators of the effective waves.

Two points are especially important here. The lumination of shadow points is zero, while the lumination of focal points is considered full.

7.7 Core Operations on Spectra

Waves obey all set operations.

7.7.1　Union of a spectrum

Let

$$S = \{\, w_{q_1}(x, r_1), \ldots, w_{q_s}(x, r_s) \,\}$$

be a finite wave spectrum with combined period T_c.

Set-level description.　On a finite domain $[s, e]$, the union selects integers hit by at least one wave:

$$\boxed{\;\bigcup_{i=1}^{s} w_{q_i}(x, r_i, s, e)\;}$$

Union as a single wave.　Over one combined period, the union spectrum reduces to a single wave with a phase set:

$$\boxed{\;\bigcup_{i=1}^{s} w_{q_i}(x, r_i) = w_{T_c}(x, M).\;}$$

Phase construction.　The phase set $M \subseteq \{0, 1, \ldots, T_c - 1\}$ is obtained by scanning one period and collecting integer roots of the product wave:

$$\boxed{\;M = iroot\!\left(\prod_{i=1}^{s} v_{q_i}(x, r_i, 0, T_c - 1)\right)\;}$$

Calculating a union is straightforward. Also, since all roots are integers here, the *iroot* operator is effectively redundant.

Example.

$$w_3(x, 1) \cup w_5(x, 4) = w_{15}\!\left(x, \{1, 4, 7, 10, 13\} \cup \{4, 9, 14\}\right) = w_{15}\!\left(x, \{1, 4, 7, 9, 10, 13, 14\}\right).$$

7.7.2　Intersection of a spectrum and the focal point

On a finite domain,

$$\boxed{\;\bigcap_{i=1}^{s} w_{q_i}(x, r_i, s, e)\;}$$

Focal point.　Let T_c be the combined period. The **focal point** of the spectrum is the smallest nonnegative integer where all waves meet:

$$\boxed{\;f = \mathrm{focal}(S) \in [0, T_c) \quad \text{such that} \quad f \in w_{q_i}(x, r_i) \text{ for all } i.\;}$$

Sum-wave characterization. Define the sum wave over one combined period:

$$s_S(x) = \sum_{i=1}^{s} v_{q_i}(x, r_i, 0, T_c - 1).$$

If the generators q_i are pairwise coprime, the focal point is uniquely determined by

$$f = iroot(s_S(x) = 0) \quad \text{in } [0, T_c).$$

If the generators are not coprime, a focal point may or may not exist; whenever it exists, it must satisfy

$$f \in iroot(s_S(x) = 0) \quad \text{in } [0, T_c).$$

Intersection as a wave. Whenever a periodic intersection exists (which exists for all finite combinations of prime waves), the intersection spectrum reduces to a single wave:

$$\bigcap_{i=1}^{s} w_{q_i}(x, r_i) = w_{T_c}(x, f).$$

7.7.3 Connection to CRT

The focal point f is the solution of the system of congruences:

$$x \equiv r_i \pmod{q_i} \qquad (i = 1, \ldots, s).$$

In particular, for distinct primes p_i with $T_c = \prod p_i$:

$$f \equiv \sum_{i=1}^{s} r_i \left(\tfrac{T_c}{p_i}\right)^{p_i - 1} \pmod{T_c}.$$

7.7.4 Shadow (complement) of a spectrum

$$\text{Shadow}(S) = U \setminus \text{Cross}(S) = U \setminus \left(\bigcup_{i=1}^{s} w_{q_i}(x, r_i)\right).$$

Shadow points are exactly the integers that avoid all crossings.

7.7.5 Parallel Waves

Two waves are said to be *parallel* if they have the same generator but differ only by a fundamental phase.

Parallel waves do not generate a focal point. Consequently, their intersection set is empty. The union of parallel waves equals a wave with a union of phases.

Example:

$$w_3(x, 1) \cap w_3(x, 2) = \emptyset, \qquad w_3(x, 1) \cup w_3(x, 2) = w_3(x, (1 \cup 2)) = w_3(x, \{1, 2\}).$$

7.7.6   Identical Waves

Two waves are said to be *identical* if and only if they have the same generator and the same fundamental phase.

$$w_3(x, 31) = w_3(x, 7) \quad \text{since } 31 \bmod 3 = 7 \bmod 3.$$

7.7.7   Shift on Waves

The shift of a wave by integer h is defined as:

$$\boxed{\text{shift}\big(w_T(x, r, s, e),\, h\big) = w_T(x, r+h, s+h, e+h).}$$

7.7.8   Waves: Instantaneous Knowledge Transfer

If we know the status of a wave at one point, it is equivalent to knowing the entire wave and the status of all points (because the wave is defined globally by its generator and phase). The point does not need to lie inside the displayed domain; it can be anywhere. The wave transfers knowledge instantly to all points in the whole universe.

In many real problems (like twin primes), we have information about different waves (or even the same wave) at different points. We can transfer all the knowledge to one point and formulate the whole problem.

7.8   Intersections Examples

Coprime spectra.

A **coprime spectrum** is a spectrum whose generators are pairwise coprime. A coprime spectrum has exactly **one** focal point in the fundamental period $[0, T_c)$.

$$I(x) = \bigcap \{\, w_2(x, 1),\ w_3(x, 2) \,\}$$

Here $T_c = \text{lcm}(2, 3) = 6$. Draw the sum-wave $s(x) = v_2(x, 1) + v_3(x, 2)$ and locate its integer root.

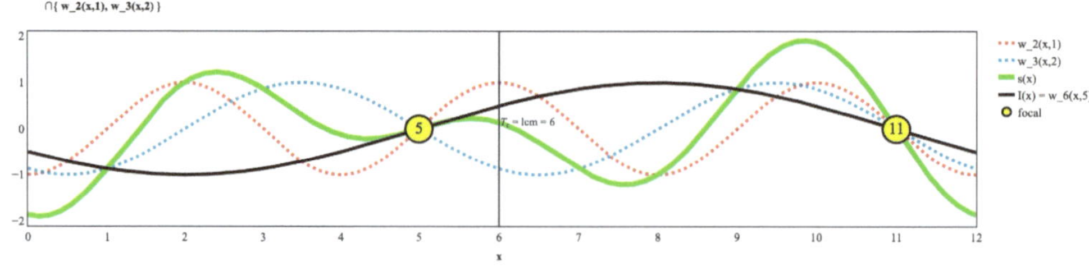

Figure 7.3

The focal point is $f = 5$, so the intersection wave is:

$$I(x) = w_6(x, 5).$$

More brief examples (as drawn).

$$\bigcap\{\, w_3(x,1),\ w_5(x,2)\,\}$$

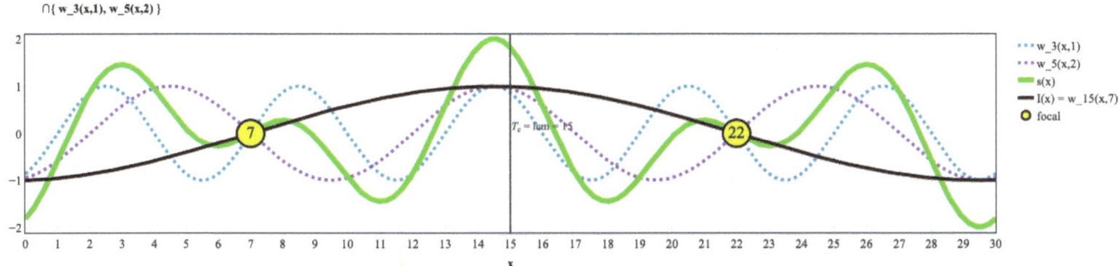

Figure 7.4

So $I(x) = w_{15}(x,7)$.

$$\bigcap\{\, w_8(x,3),\ w_9(x,2)\,\}$$

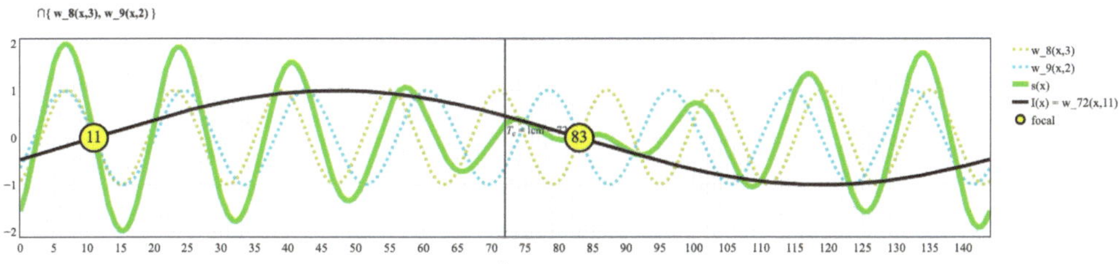

Figure 7.5

So $I(x) = w_{72}(x,11)$.

$$\bigcap\{\, w_{11}(x,5),\ w_2(x,0)\,\}$$

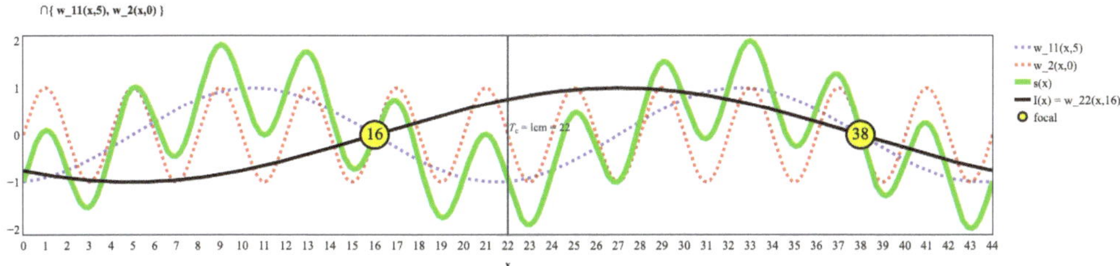

Figure 7.6

So $I(x) = w_{22}(x,16)$.

$$\bigcap\{\, w_2(x,1),\ w_3(x,1),\ w_5(x,2)\,\}$$

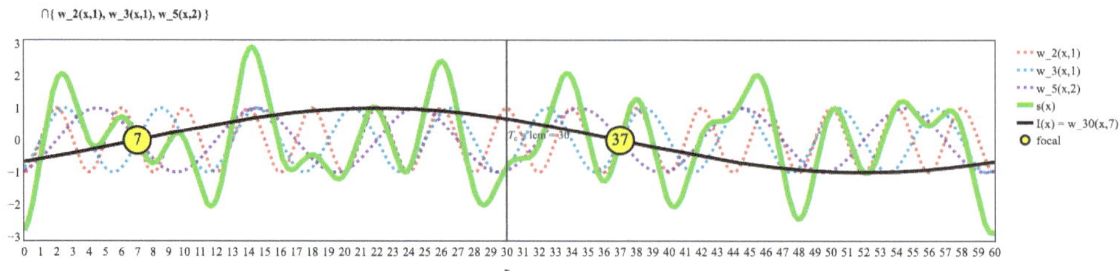

Figure 7.7:

$$\bigcap\{\, w_2(x,1),\ w_3(x,1),\ w_5(x,2)\,\}$$

So $I(x) = w_{30}(x,7)$.

When the spectrum is not coprime (as drawn).

$$\bigcap\{\, w_6(x,1),\ w_3(x,1)\,\}$$

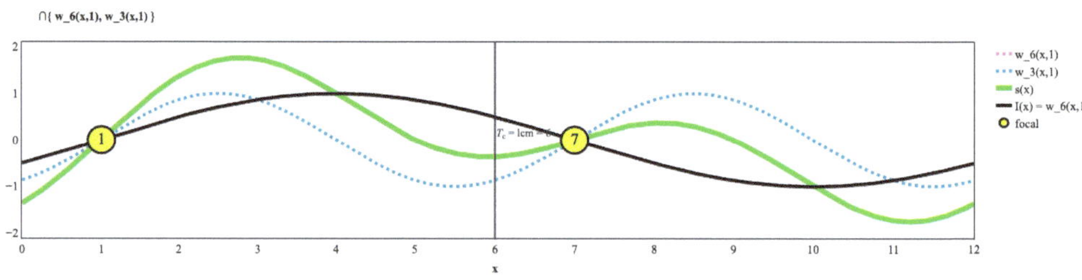

Figure 7.8

The valid intersection is $I(x) = w_6(x,1)$.

$$\bigcap\{\, w_6(x,1),\ w_3(x,0)\,\}$$

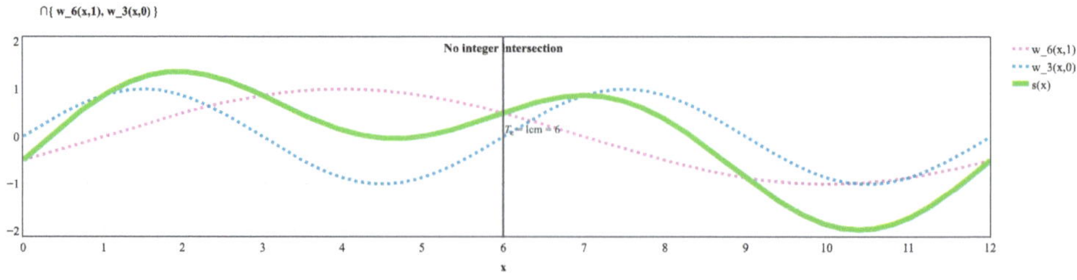

Figure 7.9

Here there is *no* integer intersection.

We can prove it using set operations, together with the fact that parallel waves do not form a focal point.

$$w_6(x, 1) \cap w_3(x, 0) = w_6(x, 1) \cap \big(w_6(x, 0) \cup w_6(x, 3)\big)$$
$$= \big(w_6(x, 1) \cap w_6(x, 0)\big) \cup \big(w_6(x, 1) \cap w_6(x, 3)\big)$$
$$= \emptyset \cup \emptyset$$
$$= \emptyset.$$

7.9 Shadow and De Morgan Laws on Waves

It is not easy to describe De Morgan in plain words. We often have different routes to the same destination. De Morgan is like a wormhole: it lets us switch to a parallel world, solve the problem there, and come back.

7.9.1 Shadow of a wave and a spectrum

Shadow as complement. Let U be the universe (here $U = \mathbb{N}_0$). For any set $A \subseteq U$,

$$\boxed{\text{Shadow}(A) = \neg A = U \setminus A.}$$

Shadow of a wave. For a wave $w_T(x, R)$ (where $R \subseteq \{0, 1, \ldots, T - 1\}$ is a residue set),

$$\boxed{w_T(x, R) = \{\, n \in U : n \bmod T \in R \,\}.}$$

Therefore the shadow wave is

$$\boxed{\text{Shadow}\big(w_T(x, R)\big) = U \setminus w_T(x, R) = w_T\big(x, R^c\big), \qquad R^c := \{0, 1, \ldots, T - 1\} \setminus R.}$$

Shadow of a spectrum. If S is a spectrum, then

$$\boxed{\text{Shadow}(S) = \neg S = U \setminus \text{Cross}(S) = U \setminus \left(\bigcup_{w \in S} w\right).}$$

7.9.2 De Morgan on Spectrum

These identities hold for any waves (or spectra) viewed as subsets of the same universe U.

Union form.

$$\boxed{w_{T_1}(x, R_1) \cup w_{T_2}(x, R_2) = \neg\Big(\neg\big(w_{T_1}(x, R_1)\big) \cap \neg\big(w_{T_2}(x, R_2)\big)\Big).}$$

Intersection form.

$$\boxed{w_{T_1}(x, R_1) \cap w_{T_2}(x, R_2) = \neg\Big(\neg\big(w_{T_1}(x, R_1)\big) \cup \neg\big(w_{T_2}(x, R_2)\big)\Big).}$$

General spectrum form. For a finite spectrum $S = \{w_1, \dots, w_s\}$,

$$\neg\left(\bigcup_{i=1}^{s} w_i\right) = \bigcap_{i=1}^{s} \neg(w_i), \qquad \neg\left(\bigcap_{i=1}^{s} w_i\right) = \bigcup_{i=1}^{s} \neg(w_i).$$

We now have enough elementary tools to build this world and to attack the most famous prime problems. You are ready to move to the next chapter.

However, before we continue, we make one broad generalization that will be used repeatedly in future seasons.

7.9.3   Spectrum period

Combined period. If a spectrum has finitely many waves with generators $T_1, \dots, T_s$, its crossing structure is periodic with a combined period. So the **combined period** (spectrum period) is

$$T_c = \operatorname{lcm}(T_1, \dots, T_s).$$

For distinct primes $q_1, \dots, q_s$,

$$T_c = q_1 q_2 \cdots q_s.$$

One-period universe. In this subsection, to talk about a single period and a shadow point at 0, we use the one-period universe

$$U_{T_c} := \{0, 1, 2, \dots, T_c - 1\},$$

and all complements $\neg(\cdot)$ are taken inside U_{T_c}.

Saturated Spectrum. A **saturated spectrum** is a spectrum for which, from some point onward toward infinity, every point is crossed by at least one wave. (Informally: beyond a threshold there are no shadow points.)

$$S = \{\, w_1(x, 0, 10, \infty) \,\}.$$

Complete Spectrum. A **complete spectrum** is a spectrum in which every point of the universe is crossed by at least one wave.

$$C = \{\, w_3(x, 0), \ w_3(x, 1), \ w_3(x, 2) \,\}.$$

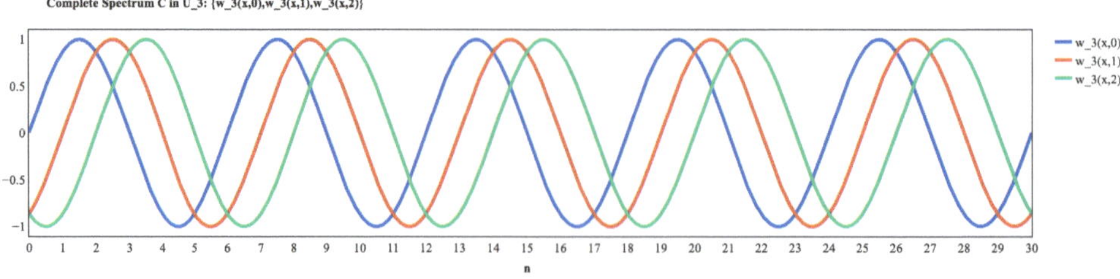

Figure 7.10: Complete spectrum $C = \{\, w_3(x, 0), \ w_3(x, 1), \ w_3(x, 2) \,\}$.

Full Spectrum. A **full spectrum** is the shadow of the empty spectrum. Equivalently, it is the spectrum in which every effective generator appears with *all* phases.

In a full prime spectrum, the number of shadow points is 0, and the lumination at each point equals the size of the generator set (namely $\pi(\sqrt{n})$).

$$F = \{w_2(x,0), w_2(x,1),$$
$$w_3(x,0), w_3(x,1), w_3(x,2),$$
$$w_5(x,0), w_5(x,1), w_5(x,2), w_5(x,3), w_5(x,4)\}.$$

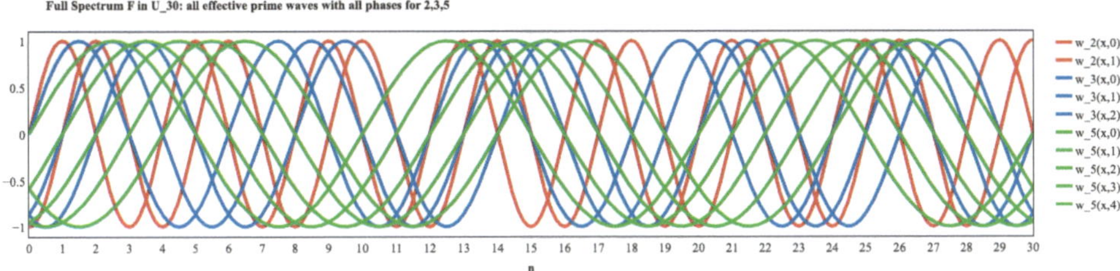

Figure 7.11: Full spectrum generated by w_2, w_3, w_5.

The lumination of all points are exactly $\pi(\sqrt{n})$ where $n = 30$

Deleting phases and counting shadow points in one period. To create shadow points, we must delete some waves. Let d_2, d_3, d_5 be the numbers of deleted phases from generators $2, 3, 5$, respectively. Then the number of shadow points in one period $T_c = 2 \cdot 3 \cdot 5 = 30$ is

$$\boxed{\#\text{shadows} = d_2\, d_3\, d_5.}$$

The first shadow point appears only when we delete at least one phase from *each* generator. For example, delete the waves whose phase equals 7 (modulo each generator):

$$D = \{w_2(x, 7 \bmod 2),$$
$$w_3(x, 7 \bmod 3),$$
$$w_5(x, 7 \bmod 5)\}.$$

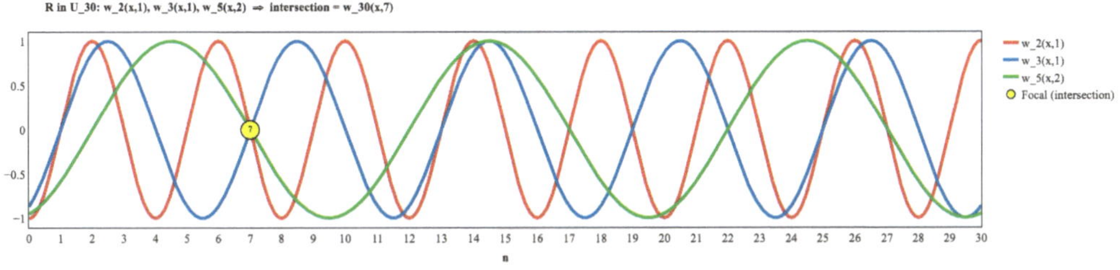

Figure 7.12: Deleted waves $w_2(x,1), w_3(x,1), w_5(x,2)$.

$$G = F \setminus D = \{w_2(x,0),$$
$$w_3(x,0), w_3(x,2),$$
$$w_5(x,0), w_5(x,1), w_5(x,3), w_5(x,4)\}.$$

This is equivalent to

$$G = w_{30}(x, R_{30} \setminus \{7\}) = \neg\, w_{30}(x,7).$$

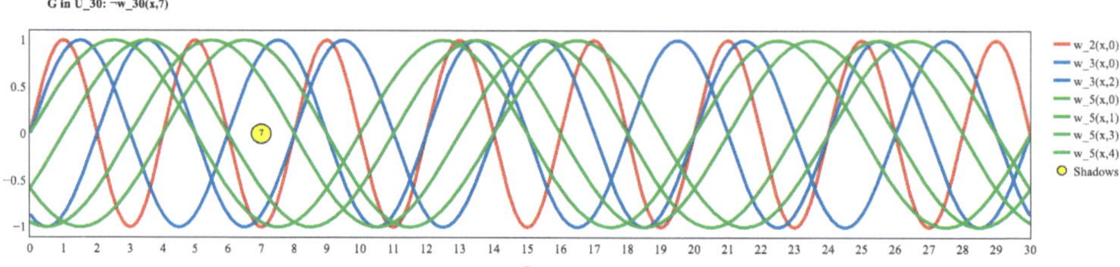

Figure 7.13: Remaining waves after deleting $w_2(x,1), w_3(x,1), w_5(x,2)$.

This chapter is the most important chapter of the book, and the Number–Wave Framework is the core foundation of our discoveries. Make sure you understand the waves conceptually.

Chapter 8

Prime Filters

"Imagination is more important than knowledge. For knowledge is limited, whereas imagination embraces the entire world, stimulating progress, giving birth to evolution."
— *Albert Einstein*

Figure 8.1: Eratosthenes of Cyrene (c. 276–194 BCE)

8.1 Preface

8.1.1 Introduction

The goal of this chapter is to formalize two fundamental spectra in the Number–Wave Framework:

- **Prime Filter Spectrum (PFS)**: a spectrum that marks points already known to be *composite* because they share at least one prime factor among $\{p_1, \ldots, p_s\}$.

- **Prime Candidates Spectrum (PCS)**: the complementary spectrum, a structured *superset* of primes consisting of points that avoid all prime factors $\leq p_s$.

In this sense,

$$\mathrm{PFS}(x, p_s) \subseteq \{\text{composites}\}, \qquad \{\text{primes}\} \subseteq \mathrm{PCS}(x, p_s) \cup \{p_1, \ldots, p_s\}.$$

The chapter then shows how candidate spectra can be generated, updated from one level to the next, and used as an efficient search space for prime generation. We also build the analogous construction for twin primes.

We dedicate this chapter to Eratosthenes and Paul Pritchard—Eratosthenes for inventing the sieve and the philosophy of filtering by elimination, and Pritchard for modern wheel-sieve innovations that make those filters roll faster and farther in computational prime discovery.

Eratosthenes

Eratosthenes of Cyrene (c. 276–194 BCE) gave mathematics one of its most enduring ideas: a way to find primes not by hunting for them one by one, but by removing everything that cannot survive. The Sieve of Eratosthenes is simple enough to teach to a child and deep enough to remain a foundation of computation more than two thousand years later. Its power is not only technical but philosophical: primes are revealed by structure. Instead of forcing numbers into patterns, Eratosthenes showed that one can clear away the composites and let the irreducible numbers remain. In that sense, he was not only a calculator of primes, but one of the earliest architects of systematic mathematical filtering. His genius, however, was far wider than number theory. He was a geometer, astronomer, geographer, chronologist, poet, and the chief librarian of Alexandria—a mind equally comfortable with abstract structure and the physical world.

That broader genius appears most vividly in his work on the Earth itself. Eratosthenes famously estimated the circumference of the Earth with astonishing insight, using shadows, distance, and geometry, and he transformed geography into a quantitative science. His world map was one of the great intellectual achievements of antiquity: not merely a sketch of lands, but an attempt to organize the known world through measurement, proportion, and reason. For its time, it was a masterpiece of scientific mapping, and its spirit of measured geography shaped cartographic thought for many centuries. In Eratosthenes, we see a rare kind of thinker: a man who could sieve the integers for primes and, with the same clarity of mind, measure the planet and map the world.

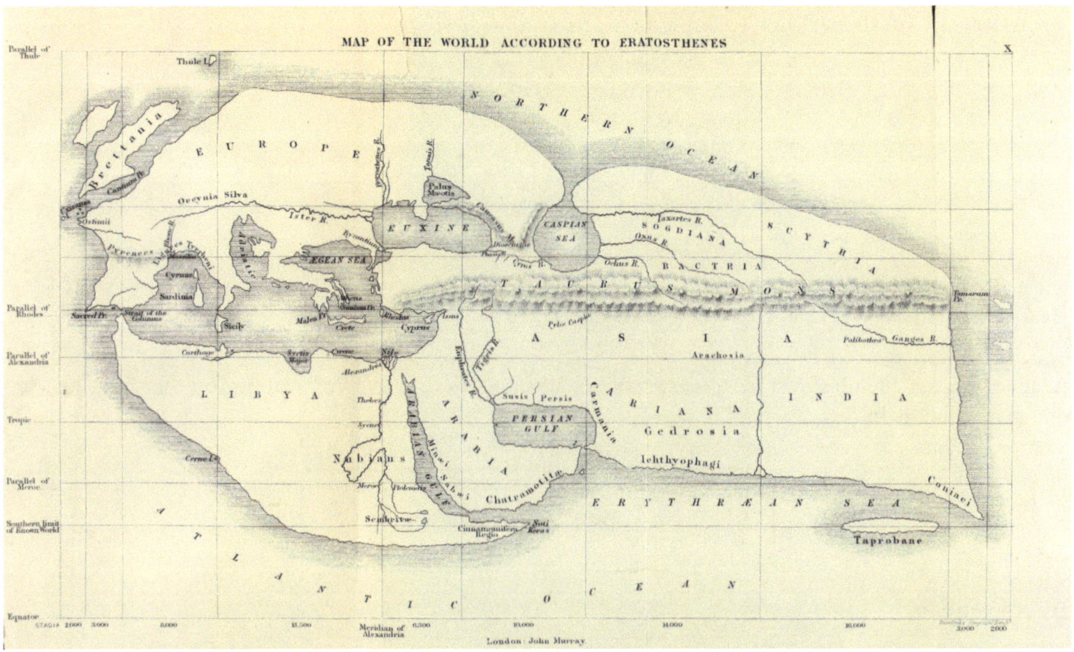

Figure 8.2: World map attributed to Eratosthenes of Cyrene.

Paul Pritchard: Making the Wheel Roll

Paul Pritchard advanced the sieve tradition into a modern, algorithmic art by refining one of its deepest practical ideas: once you know small primes, you should not waste time stepping on their multiples again. His work on wheel-based, dynamic sieving turns that insight into a living mechanism—patterns are reused, the search space shrinks intelligently, and prime generation becomes something that can be updated and streamed rather than recomputed from scratch. In the spirit of this chapter, Pritchard represents the second leap after Eratosthenes: not only filtering composites, but engineering the filter so it scales—cleanly, efficiently, and with the elegance of a machine that learns its own rhythm.

Naming and Conventions

Let $p_1 = 2 < p_2 = 3 < \cdots < p_s$ be the primes and define the primorial

$$p_s\# = \prod_{i=1}^{s} p_i$$

as the combined period at level p_s. Unless an interval $[a, b]$ is explicitly written, every spectrum in this chapter is understood on the fundamental period

$$[1, p_s\#].$$

We write $\neg(\cdot)$ for the shadow (complement) *inside the same fundamental period.* Abbreviations:

$$\begin{aligned}
\text{PFS} &= \text{Prime Filter Spectrum,} \\
\text{PCS} &= \text{Prime Candidates Spectrum,} \\
\text{TPFS} &= \text{Twin Prime Filter Spectrum,} \\
\text{TPCS} &= \text{Twin Prime Candidates Spectrum.}
\end{aligned}$$

8.2 Prime Filter Spectrum (PFS)

At level p_s, the *Prime Filter Spectrum* crosses all integers in one primorial period that have at least one prime factor in $\{p_1,\ldots,p_s\}$:

$$\boxed{\text{PFS}(x,p_s,1,p_s\#) = \bigcup_{i=1}^{s} w_{p_i}(x,0,1,p_s\#).}$$

By convention,

$$\boxed{\text{PFS}(x,p_s) = \text{PFS}(x,p_s,1,p_s\#).}$$

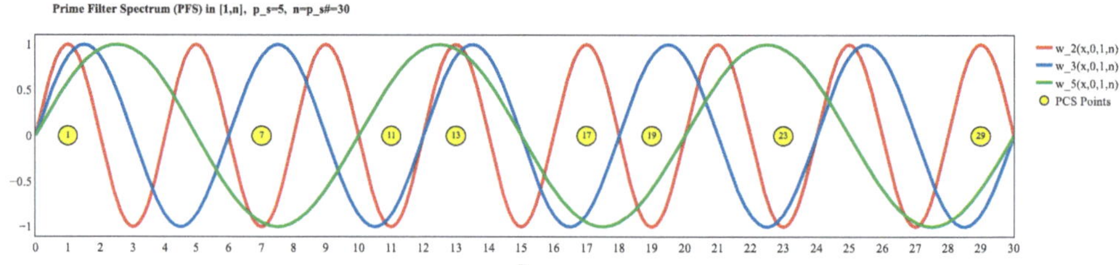

Figure 8.3: PFS(x,5)

Counting filtered points (Inclusion–Exclusion)

Inside one primorial period, the count follows directly from Inclusion–Exclusion:

$$\boxed{\left|\text{PFS}(x,p_s)\right| = p_s\# \sum_{\emptyset \neq J \subseteq \{1,\ldots,s\}} (-1)^{|J|+1} \frac{1}{\prod_{j\in J} p_j}.}$$

Example: PFS$(x,5)$.   Here $p_s = 5$ and $p_s\# = 30$. Then

$$\left|\text{PFS}(x,5)\right| = 30 \left[\left(\frac{1}{2}+\frac{1}{3}+\frac{1}{5}\right) - \left(\frac{1}{6}+\frac{1}{10}+\frac{1}{15}\right) + \frac{1}{30} \right]$$
$$= (15+10+6) - (5+3+2) + 1 = 22.$$

8.3　Prime Candidates Spectrum (PCS)

The *Prime Candidates Spectrum* is the shadow of the filter spectrum inside one primorial period:

$$\mathrm{PCS}(x, p_s) = \neg\, \mathrm{PFS}(x, p_s).$$

By De Morgan on waves, it also has the intersection form

$$\mathrm{PCS}(x, p_s) = \bigcap_{i=1}^{s} \neg\, w_{p_i}(x, 0, 1, p_s\#).$$

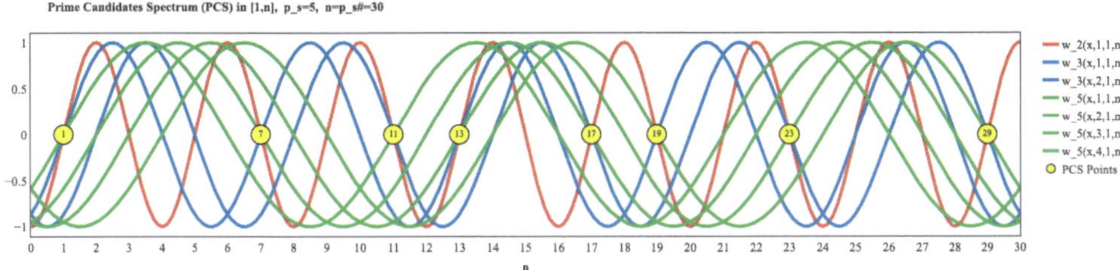

Figure 8.4: PCS(x,5)

Fact.　A point in $\mathrm{PCS}(x, p_s)$ has no prime factor $\leq p_s$.

Counting candidates in one period

Since the generators $p_1, \ldots, p_s$ are pairwise coprime, the candidate count in one primorial period is

$$\left|\mathrm{PCS}(x, p_s)\right| = p_s\# \prod_{i=1}^{s}\left(1 - \frac{1}{p_i}\right).$$

Equivalently (because $p_1 = 2$),

$$\left|\mathrm{PCS}(x, p_s)\right| = \prod_{i=2}^{s}(p_i - 1).$$

Also, inside one period,

$$\left|\mathrm{PCS}(x, p_s)\right| = p_s\# - \left|\mathrm{PFS}(x, p_s)\right|.$$

8.3.1　Guess a Prime Number

If we must guess primes quickly, betting on prime candidates is the best strategy.

Game.　You have five minutes to write down 25 integers larger than 2,500,000. The winner lists the most primes.

Setup. Compute primorials until you exceed 2,500,000:

$$2 \cdot 3 \cdot 5 \cdot 7 \cdot 11 \cdot 13 \cdot 17 = 510{,}510, \qquad 2 \cdot 3 \cdot 5 \cdot 7 \cdot 11 \cdot 13 \cdot 17 \cdot 19 = 9{,}699{,}690.$$

Thus $s = 7$, $p_s = 17$, and $p_s\# = 510{,}510$. We need k such that

$$k \cdot p_s\# > 2{,}500{,}000 \quad \Rightarrow \quad k > \frac{2{,}500{,}000}{510{,}510} \quad \Rightarrow \quad k \geq 5.$$

Take $k = 5$, so the base is

$$B = 5 \cdot 510{,}510 = 2{,}552{,}550.$$

Pick residues. Any residue $r \in \mathrm{PCS}(x, 17)$ produces a candidate $N = B + r$. A simple choice is to take residues that are primes > 17 (and also $r = 1$):

$$R = [1, 19, 23, 29, 31, 37, 41, 43, 47, 53, 59, 61, 67, 71, 73, 79, 83, 89, 97, 101, 103, 107, 109, 113, 127].$$

Then each candidate $N = B + r$ has no prime factor ≤ 17.

Candidates.

Base B	Residue $r \in R$	Candidate $N = B + r$	Classification
2,552,550	1	2,552,551	*Composite*
2,552,550	19	2,552,569	*Composite*
2,552,550	23	2,552,573	*Composite*
2,552,550	29	2,552,579	*Composite*
2,552,550	31	2,552,581	*Prime*
2,552,550	37	2,552,587	*Prime*
2,552,550	41	2,552,591	*Composite*
2,552,550	43	2,552,593	*Composite*
2,552,550	47	2,552,597	*Prime*
2,552,550	53	2,552,603	*Prime*
2,552,550	59	2,552,609	*Composite*
2,552,550	61	2,552,611	*Prime*
2,552,550	67	2,552,617	*Composite*
2,552,550	71	2,552,621	*Prime*
2,552,550	73	2,552,623	*Prime*
2,552,550	79	2,552,629	*Prime*
2,552,550	83	2,552,633	*Composite*
2,552,550	89	2,552,639	*Composite*
2,552,550	97	2,552,647	*Prime*
2,552,550	101	2,552,651	*Prime*
2,552,550	103	2,552,653	*Prime*
2,552,550	107	2,552,657	*Prime*
2,552,550	109	2,552,659	*Prime*
2,552,550	113	2,552,663	*Composite*
2,552,550	127	2,552,677	*Prime*

These are 25 prime candidates $> 2.5 \times 10^6$ generated from $\mathrm{PCS}(x, 17)$.

8.4 Generating Prime Candidates Incrementally

Going from level p_s to level p_{s+1} is a two-step construction.

Step 1 (stack p_{s+1} blocks). Copy the previous candidate pattern across p_{s+1} consecutive blocks of length $p_s\#$:

$$\bigcup_{k=0}^{p_{s+1}-1} \text{PCS}\big(x, p_s,\, k \cdot p_s\# + 1,\, (k+1) \cdot p_s\#\big).$$

Step 2 (remove the new filter). Subtract the new prime wave $w_{p_{s+1}}(x, 0, 1, p_{s+1}\#)$:

$$\text{PCS}(x, p_{s+1}, 1, p_{s+1}\#) = \left(\bigcup_{k=0}^{p_{s+1}-1} \text{PCS}\big(x, p_s,\, k \cdot p_s\# + 1,\, (k+1) \cdot p_s\#\big) \right) - w_{p_{s+1}}(x, 0, 1, p_{s+1}\#).$$

Example: build $\text{PCS}(x, 5, 1, 30)$ from $\text{PCS}(x, 3, 1, 6)$. Here $p_s = 3$, $p_s\# = 6$, $p_{s+1} = 5$, $p_{s+1}\# = 30$. We know

$$\text{PCS}(x, 3, 1, 6) = \{1, 5\}.$$

Stack five blocks:

$$\{1, 5\} \cup \{7, 11\} \cup \{13, 17\} \cup \{19, 23\} \cup \{25, 29\}.$$

Now subtract the new filter wave

$$w_5(x, 0, 1, 30) = \{5, 10, 15, 20, 25, 30\}.$$

Therefore

$$\boxed{\text{PCS}(x, 5, 1, 30) = \{1, 7, 11, 13, 17, 19, 23, 29\}.}$$

8.5 Prime Generation Using Prime Filters

Prime candidates still include higher-order composites. The remaining task is to cross out crossings produced by higher-order prime waves.

Higher-order waves. At level p_s, for a number n, the higher-order primes are

$$p_{s+1} \leq q \leq \sqrt{n},$$

and their waves are the only waves capable of crossing out elements of $\text{PCS}(x, p_s)$ below n.

Example (primes up to 200). Find the largest primorial below 200:

$$5\# = 30, \qquad 7\# = 210.$$

Hence $s = 3$ and the level prime is $p_s = 5$. The effective primes are

$$L(200) = \{2, 3, 5\}.$$

The candidate residues in one period are

$$\text{PCS}(x, 5, 1, 30) = \{1, 7, 11, 13, 17, 19, 23, 29\}.$$

All primes < 200 are either in $L(200)$ or of the form $k \cdot 30 + r$ with $r \in \mathrm{PCS}(x, 5, 1, 30)$. Thus the prime *candidates* under 200 are

$$C(200) = \{1, 7, 11, 13, 17, 19, 23, 29, 31, 37, 41, 43, 47, 49, 53, 59, 61, 67, 71, 73, 79, 83,$$
$$89, 97, 101, 103, 107, 109, 113, 119, 121, 127, 131, 133, 137, 139, 143, 149,$$
$$151, 157, 161, 163, 167, 169, 173, 179, 181, 187, 191, 193, 197, 199\}.$$

Now cross out higher-order composites under 200. Since $\sqrt{200} < 15$, only the higher-order primes $7, 11, 13$ can produce new crossings via products.

	7	11	13
7	49		
11	77	121	
13	91	143	169
17	119	187	289
19	133	209	
23	161		
29	203		

We stop when the product exceeds 200. Therefore the higher-order composites below 200 are

$$H(200) = \{49, 77, 91, 119, 121, 133, 143, 161, 169, 187\}.$$

Collecting everything:

$$L(200) = \{2, 3, 5\}, \qquad P(200) = (L(200) \cup C(200)) - H(200) - \{1\}.$$

Therefore

$$P(200) = \{2, 3, 5, 7, 11, 13, 17, 19, 23, 29, 31, 37, 41, 43, 47, 53, 59, 61, 67, 71, 73, 79, 83,$$
$$89, 97, 101, 103, 107, 109, 113, 127, 131, 137, 139, 149, 151, 157, 163, 167, 173,$$
$$179, 181, 191, 193, 197, 199\}.$$

Since $L(200)$, $C(200)$, and $H(200)$ are pairwise disjoint,

$$\boxed{\pi(200) = |P(200)| = |L(200)| + |C(200)| - |H(200)| - 1.}$$

8.6   Reconstructing Filters from Known Primes

Prime generation can resume from a checkpoint: if primes are already known up to a bound, we can rebuild the relevant candidate spectrum.

Example: rebuild $\mathrm{PCS}(x, 5, 1, 30)$ from $P(70)$.

Let

$$P(70) = \{2, 3, 5, 7, 11, 13, 17, 19, 23, 29, 31, 37, 41, 43, 47, 53, 59, 61, 67\}.$$

Decompose 70 by the primorial $5\# = 30$:

$$70 = 2 \cdot 30 + 10, \qquad s = 3, \quad p_s = 5, \quad L(70) = \{2, 3, 5\}.$$

Higher-order primes (beyond 5) that are known from $P(70)$ are

$$H_s = \{7, 11, 13, 17, 19, 23, 29\}.$$

To rebuild $\mathrm{PCS}(x, 5, 1, 30)$, test crossings inside one period using products of higher-order primes. Here $7 \cdot 7 = 49 > 30$, so no new crossing appears inside $[1, 30]$. Hence

$$\boxed{\mathrm{PCS}(x, 5, 1, 30) = \{1, 7, 11, 13, 17, 19, 23, 29\}.}$$

With $\mathrm{PCS}(x, 5)$ reconstructed, we continue the same block-by-block procedure.

Example: Reconstructing $\mathrm{PCS}(x, 7, 1, 210)$ (one full primorial period)

Here $p_s = 7$ and the fundamental period is the primorial

$$p_s\# = 7\# = 2 \cdot 3 \cdot 5 \cdot 7 = 210.$$

Recall that $\mathrm{PCS}(x, 7, 1, 210)$ is the set of integers in $[1, 210]$ that have *no* prime factor in $\{2, 3, 5, 7\}$. Equivalently, it is the full set of residues $r \in [1, 210]$ with $\gcd(r, 210) = 1$.

Known primes under 210.

$$\begin{aligned}
P(210) = \{&2, 3, 5, 7, 11, 13, 17, 19, 23, 29, 31, 37, 41, 43, 47, 53, 59, 61, 67, 71, \\
&73, 79, 83, 89, 97, 101, 103, 107, 109, 113, 127, 131, 137, 139, 149, 151, \\
&157, 163, 167, 173, 179, 181, 191, 193, 197, 199\}.
\end{aligned}$$

Level primes and higher-order primes. At level $p_s = 7$, the level primes are $L = \{2, 3, 5, 7\}$. The higher-order primes inside the period are

$$\begin{aligned}
H = P(210) \setminus L = \{&11, 13, 17, 19, 23, 29, 31, 37, 41, 43, 47, 53, 59, 61, 67, 71, \\
&73, 79, 83, 89, 97, 101, 103, 107, 109, 113, 127, 131, 137, 139, \\
&149, 151, 157, 163, 167, 173, 179, 181, 191, 193, 197, 199\}.
\end{aligned}$$

Higher-order composites that still live in $\mathrm{PCS}(x, 7)$. Numbers in $\mathrm{PCS}(x, 7, 1, 210)$ may still be composite; the only composites ≤ 210 with no factor in $\{2, 3, 5, 7\}$ are products of higher-order primes that remain ≤ 210. Since 11 is the smallest element of H, we only need to check products starting at 11:

$$\begin{aligned}
\text{Products from } H \text{ under } 210 = \{&11 \cdot 11 = 121, \ 11 \cdot 13 = 143, \ 11 \cdot 17 = 187, \\
&11 \cdot 19 = 209, \ 13 \cdot 13 = 169\}.
\end{aligned}$$

Therefore, the full candidate spectrum in one period is

$$\begin{aligned}
\mathrm{PCS}(x, 7, 1, 210) = \{&1, 11, 13, 17, 19, 23, 29, 31, 37, 41, 43, 47, 53, 59, 61, 67, 71, 73, \\
&79, 83, 89, 97, 101, 103, 107, 109, 113, 121, 127, 131, 137, 139, 143, 149, \\
&151, 157, 163, 167, 169, 173, 179, 181, 187, 191, 193, 197, 199, 209\}.
\end{aligned}$$

Sanity check (count). Using the general count $\left|\mathrm{PCS}(x, p_s)\right| = \prod_{i=2}^{s}(p_i - 1)$, we get

$$\left|\mathrm{PCS}(x, 7)\right| = (3 - 1)(5 - 1)(7 - 1) = 2 \cdot 4 \cdot 6 = 48,$$

which matches the 48 residues listed above.

Symmetry. Because $\gcd(r, 210) = 1 \iff \gcd(210 - r, 210) = 1$, the set is symmetric around $210/2 = 105$: if $r \in \mathrm{PCS}(x, 7, 1, 210)$ then $210 - r \in \mathrm{PCS}(x, 7, 1, 210)$.

8.7 Twin Prime Filters Spectrum (TPFS)

Twin primes are handled by the same logic, but with two phases $\{0, 2\}$.

At level p_s, the twin filter spectrum is

$$\mathrm{TPFS}(x, p_s) = \bigcup_{i=1}^{s} w_{p_i}\big(x, \{0, 2\}, 0, p_s\#\big).$$

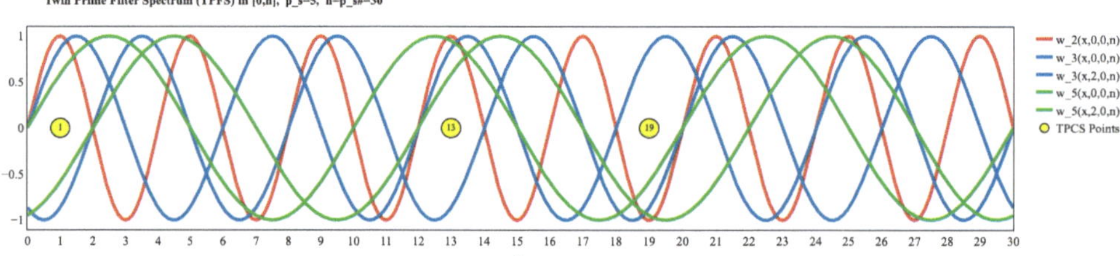

Figure 8.5: TPFS(x,5)

8.7.1 Twin Prime Candidates Spectrum (TPCS)

The twin candidates are the shadow of TPFS:

$$\mathrm{TPCS}(x, p_s) = \neg\,\mathrm{TPFS}(x, p_s).$$

Equivalently,

$$\mathrm{TPCS}(x, p_s) = \bigcap_{i=1}^{s} \neg\, w_{p_i}\big(x, \{0, 2\}, 0, p_s\#\big).$$

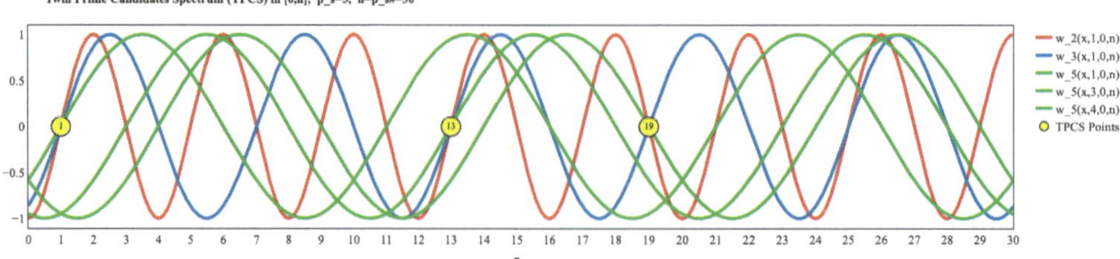

Figure 8.6: TPCS(x,5)

Counting twin candidates in one period

$$\left| \text{TPCS}(x, p_s) \right| = p_s \# \cdot \frac{1}{2} \prod_{3 \leq p_i \leq p_s} \left(1 - \frac{2}{p_i} \right)$$

and using $p_s \# = 2 \cdot \prod_{3 \leq p_i \leq p_s} p_i$, this simplifies to

$$\left| \text{TPCS}(x, p_s) \right| = \prod_{3 \leq p_i \leq p_s} (p_i - 2).$$

Example: $p_s = 5$

Here $p_s \# = 2 \cdot 3 \cdot 5 = 30$. The count gives

$$\left| \text{TPCS}(x, 5) \right| = (3 - 2)(5 - 2) = 3.$$

Solving in one period yields

$$\text{TPCS}(x, 5, 1, 30) = \{1, 13, 19\}.$$

So for all sufficiently large upper points q, every twin prime $(q - 2, q)$ must satisfy

$$q \equiv 1, 13, 19 \pmod{30}.$$

Prime Candidate Spectrum is a valuable spectrum, but the growth rate of it is still more than prime numbers. So primes are vanishing in Prime Candidates. Its strength is that it is *periodic* and *cheap*: once $\text{PCS}(x, p_s)$ is built on $[1, p_s\#]$, it repeats across the number line by shifts of $p_s\#$. However, primes are not periodic objects, so reaching the Prime Spectrum requires additional, more dynamic spectra that react to the growth of n and to higher-order crossings.

In the next chapter we move from PCS to the spectra that capture the actual mechanics of prime formation in this framework, and we prepare the same road for all other prime patterns. In the final chapter, we return to twin primes.

Chapter 9

Prime Numbers

"Mathematicians have tried in vain to this day to discover some order in the sequence of prime numbers, and we have reason to believe that it is a mystery into which the human mind will never penetrate."

— Leonhard Euler

Figure 9.1: Leonhard Euler (1707–1783).

9.1 Preface

Leonhard Euler: The King of Numbers

Leonhard Euler stands at the center of mathematics like a king—not because he claimed a throne, but because so much of modern mathematics still speaks his language. With fearless imagination and unmatched technical power, he built bridges between arithmetic, geometry, analysis, and the emerging physics of his era, turning scattered problems into unified methods. In number theory, he did not treat primes as isolated curiosities; he treated them as the atoms of a vast arithmetic structure. He showed mathematicians how infinite products, series, and generating ideas could become precise tools rather than philosophical dreams. Euler's work does more than solve problems. It sets a standard for mathematical thinking itself: bold, elegant, and endlessly productive.

We dedicate this chapter to Leonhard Euler—whose vision transformed primes from isolated curiosities into the atoms of a grand arithmetic structure, and whose products, series, and infinite methods still shape the language through which we explore the kingdom of numbers.

9.2 Overview

We have already established the Number–Wave Framework. In this chapter, we turn to prime numbers. Our goal is to design prime spectra and rewrite prime counting in the language of waves, shadows, and dependency factors.

9.3 Prime Effective Spectrum (PES)

Definition (wave form). The **Prime Effective Spectrum** is the union of all effective prime waves with phase 0 on the domain $[1, n]$:

$$\mathrm{PES}(x, n) = \bigcup_{i=1}^{s} w_{p_i}(x, 0, 1, n).$$

Each wave $w_{p_i}(x, 0, 1, n)$ crosses all multiples of p_i in $[1, n]$. Therefore, $\mathrm{PES}(x, n)$ is exactly the set of integers in $[1, n]$ that are divisible by at least one effective prime.

The effective primes are

$$p_1 < p_2 < \cdots < p_s \leq \sqrt{n}.$$

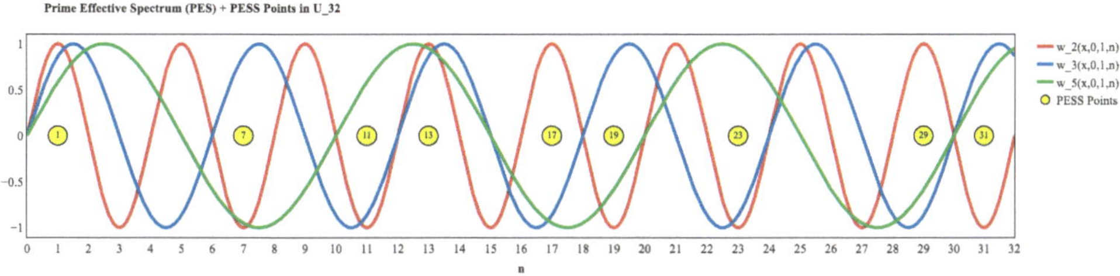

Figure 9.2: $\mathrm{PES}(x, 32)$.

Counting $|\mathrm{PES}(x,n)|$ **by inclusion–exclusion.**   From the previous chapter, the number of crossed points is

$$|\mathrm{PES}(x,n)| = \sum_{i=1}^{s} \left\lfloor \frac{n}{p_i} \right\rfloor - \sum_{1 \le i < j \le s} \left\lfloor \frac{n}{p_i p_j} \right\rfloor + \sum_{1 \le i < j < k \le s} \left\lfloor \frac{n}{p_i p_j p_k} \right\rfloor - \cdots .$$

Equivalently,

$$|\mathrm{PES}(x,n)| = n - \big(\pi(n) - \pi(\sqrt{n}) + 1\big).$$

9.3.1   Prime Effective Shadow Spectrum (PESS)

Definition (shadow of PES).   The **Prime Effective Shadow Spectrum** is the shadow of $\mathrm{PES}(x,n)$ inside the universe U_n:

$$\mathrm{PESS}(x,n) = \neg \mathrm{PES}(x,n).$$

So, by De Morgan's law,

$$\mathrm{PESS}(x,n) = \bigcap_{i=1}^{s} w_{p_i}\big(x,\, R_{p_i} \setminus \{0\},\, 1,\, n\big).$$

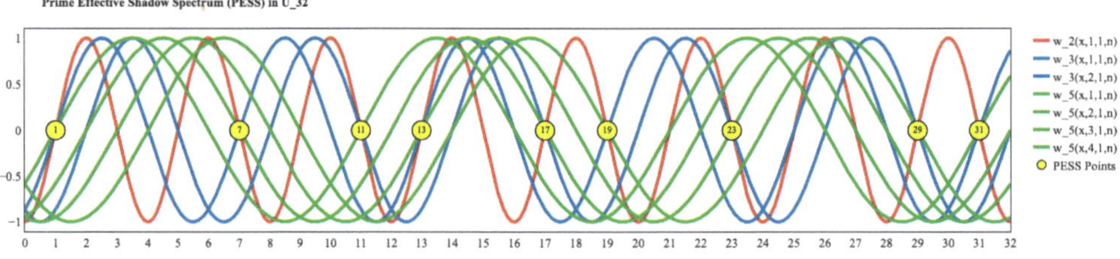

Figure 9.3: $\mathrm{PESS}(x, 32)$.

The lumination of every PESS point is equal to $\pi(\sqrt{n})$.
The size of $|\mathrm{PESS}(x,n)|$ can be determined by

$$|\mathrm{PESS}(x,n)| = \pi(n) - \pi(\sqrt{n}) + 1.$$

9.3.2   Independent Prime Effective Spectrum (IPES)

Definition (independent phases).   The **Independent Prime Effective Spectrum** replaces each deterministic phase by an unknown phase $*$:

$$\mathrm{IPES}(x,n) = \bigcup_{i=1}^{s} w_{p_i}(x, *, 1, n).$$

For each wave $w_{p_i}(x, *, 1, n)$, the independent model gives

$$|w_{p_i}(x, *, 1, n)| = \frac{n}{p_i},$$

so inclusion–exclusion now contains no floor functions.

Independent inclusion–exclusion.

$$|\text{IPES}(x,n)| = \sum_{i=1}^{s} \frac{n}{p_i} - \sum_{1 \leq i < j \leq s} \frac{n}{p_i p_j} + \sum_{1 \leq i < j < k \leq s} \frac{n}{p_i p_j p_k} - \cdots.$$

Also, by De Morgan's law,

$$\text{IPES}(x,n) = \bigcup_{i=1}^{s} w_{p_i}(x, *, 1, n) = \neg \bigcap_{i=1}^{s} w_{p_i}(x, *_{-1}, 1, n).$$

Therefore,

$$|\text{IPES}(x,n)| = n - n \prod_{p_i \leq \sqrt{n}} \left(1 - \frac{1}{p_i}\right).$$

We define the **Prime Generating Function** by

$$G_x(n) := G(x,n) := \prod_{x < p \leq n} \left(1 - \frac{x}{p}\right),$$

where $x \geq 0$ is fixed and the product runs over primes p.
So,

$$|\text{IPES}(x,n)| = n\big(1 - G(1, \sqrt{n})\big).$$

9.3.3   Independent Prime Effective Shadow Spectrum (IPESS)

Definition.   The **Independent Prime Effective Shadow Spectrum** is the shadow of $\text{IPES}(x,n)$ inside U_n:

$$\text{IPESS}(x,n) = \neg \bigcup_{i=1}^{s} w_{p_i}(x, *, 1, n) = \bigcap_{i=1}^{s} \neg w_{p_i}(x, *, 1, n) = \bigcap_{i=1}^{s} w_{p_i}(x, *_{-1}, 1, n).$$

Independent shadow size.   By standard product algebra on the independent spectrum,

$$|\text{IPESS}(x,n)| = n \prod_{i=1}^{s} \left(1 - \frac{1}{p_i}\right) = n \prod_{p_i \leq \sqrt{n}} \left(1 - \frac{1}{p_i}\right) = n\, G(1, \sqrt{n}).$$

Effective prime waves are asymptotically independent: as $n \to \infty$, they complete their periods infinitely many times, so their dependency weakens. Going forward, new waves are introduced into the models, but high-frequency prime waves such as w_2 have the dominant power in shaping the spectrum structure over long domains. So the major behaviour and structure of every spectrum are shaped by high-frequency waves. That is the main reason for convergence at infinity in the field of prime numbers.

Although effective prime waves are asymptotically independent, they are not independent on finite numbers, especially when they interact through operations and produce large periods.

Independence unlocks the algebra; at finite n, the remaining dependency appears as an error term, which we now isolate as the Total Tail Error.

9.4   Prime Counting and Total Tail Error

9.4.1   Exact identity for $\pi(n)$

Every composite $m \le n$ is divisible by some prime $\le \sqrt{m} \le \sqrt{n}$, so it is crossed by $\mathrm{PES}(x, n)$. Therefore, the shadow $\mathrm{PESS}(x, n)$ consists of:

- the integer 1,

- and all primes in the interval $(\sqrt{n}, n]$.

Hence,

$$\pi(n) = n - |\mathrm{PES}(x, n)| + \pi(\sqrt{n}) - 1 = |\mathrm{PESS}(x, n)| + \pi(\sqrt{n}) - 1.$$

We already know IPESS and IPES. So the remaining task is to understand the relation between the actual spectra and the independent spectra.

9.4.2   Total tail error $E_1(n)$

The only difference between $\mathrm{PES}(x, n)$ and $\mathrm{IPES}(x, n)$ is the presence of floor functions. We encode that discrepancy in the **total tail error**.

Definition.   For each nonempty subset $S \subseteq \{1, \ldots, s\}$, recall

$$T_S := \prod_{i \in S} p_i, \qquad n = k_S T_S + r_S, \qquad 0 \le r_S < T_S,$$

so that

$$\frac{r_S}{T_S} = \frac{n}{T_S} - \left\lfloor \frac{n}{T_S} \right\rfloor.$$

Applying inclusion–exclusion to the effective waves, we define

$$E_1(n) := |\mathrm{IPES}(x, n)| - |\mathrm{PES}(x, n)| = |\mathrm{PESS}(x, n)| - |\mathrm{IPESS}(x, n)|.$$

This matches the intuition: $E_1(n)$ is the accumulated floor discrepancy between deterministic counting and independent counting.

We obtain the explicit finite alternating sum

$$E_1(n) = + \sum_{i=1}^{s} \frac{r_{\{i\}}}{p_i} - \sum_{1 \le i < j \le s} \frac{r_{\{i,j\}}}{p_i p_j} + \sum_{1 \le i < j < k \le s} \frac{r_{\{i,j,k\}}}{p_i p_j p_k} - \cdots.$$

Here,

$$r_{\{i\}} = \mathrm{focal}\big(\{\, w_{p_i}(x, n \bmod p_i) \,\}\big),$$

$$r_{\{i,j\}} = \mathrm{focal}\big(\{\, w_{p_i}(x, r_i),\ w_{p_j}(x, r_j) \,\}\big),$$

and similarly for larger subsets. The focal point can be obtained by the CRT function.

Since all waves start from the origin, the focal-point values simplify to

$$r_{\{i\}} = n \bmod p_i,$$

$$r_{\{i,j\}} = n \bmod (p_i p_j),$$

and similarly for all larger subsets.

9.4.3 Prime counting formula in independent form

Using the independent prime effective shadow spectrum, we obtain the central identity

$$\pi(n) = n \prod_{p_i \le \sqrt{n}} \left(1 - \frac{1}{p_i}\right) + E_1(n) + \pi(\sqrt{n}) - 1 = n\,G(1, \sqrt{n}) + E_1(n) + \pi(\sqrt{n}) - 1.$$

The independent main term is the independent shadow $\text{IPESS}(x, n)$; the deterministic structure of primes is carried by the total tail error $E_1(n)$, together with the lower-order correction $\pi(\sqrt{n})$.

Example: $n = 42$

Effective primes:

$$p_1 = 2, \qquad p_2 = 3, \qquad p_3 = 5.$$

Independent shadow.

$$|\text{IPESS}(42)| = 42 \prod_{p_i \le \sqrt{42}} \left(1 - \frac{1}{p_i}\right)$$

$$= 42\left(1 - \frac{1}{2}\right)\left(1 - \frac{1}{3}\right)\left(1 - \frac{1}{5}\right)$$

$$= 42 \cdot \frac{1}{2} \cdot \frac{2}{3} \cdot \frac{4}{5} = 11.2.$$

Total tail error.

$$E_1(42) = +\frac{42 \bmod 2}{2} + \frac{42 \bmod 3}{3} + \frac{42 \bmod 5}{5}$$

$$-\frac{42 \bmod 6}{6} - \frac{42 \bmod 10}{10} - \frac{42 \bmod 15}{15}$$

$$+\frac{42 \bmod 30}{30}$$

$$= +\frac{0}{2} + \frac{0}{3} + \frac{2}{5} - \frac{0}{6} - \frac{2}{10} - \frac{12}{15} + \frac{12}{30}$$

$$= 0 + 0 + 0.4 - 0 - 0.2 - 0.8 + 0.4$$

$$= -0.2.$$

Recovering $\pi(42)$. Since $\pi(\sqrt{42}) = \pi(6) = 3$,

$$\pi(42) = |\text{IPESS}(42)| + E_1(42) + \pi(\sqrt{42}) - 1 = 11.2 - 0.2 + 3 - 1 = 13,$$

which is the true prime count.

However, calculating the total tail error $E_1(n)$ has essentially the same complexity as calculating $\text{PESS}(x, n)$ itself. It is useful in additive space, but in multiplicative space it is better for the error to be defined as a product term. Since the equality relates two entities, we now define the dependency factor to connect the independent model to the dependent model.

Before proceeding, based on the oscillating nature of the error function, we record a visible two-sided bound for $\pi(n)$ obtained by truncating the alternating sum.

9.4.4   Prime counting function bound

Figure 9.4: Pafnuty Chebyshev (1821–1894).

$$n\,G(1,\sqrt{n}) + \pi(\sqrt{n}) - 1 + \sum_{i=1}^{s} \frac{r_{\{i\}}}{p_i} - \sum_{1 \le i < j \le s} \frac{r_{\{i,j\}}}{p_i p_j}$$

$$\le \pi(n) \le$$

$$n\,G(1,\sqrt{n}) + \pi(\sqrt{n}) - 1 + \sum_{i=1}^{s} \frac{r_{\{i\}}}{p_i} - \sum_{1 \le i < j \le s} \frac{r_{\{i,j\}}}{p_i p_j} + \sum_{1 \le i < j < k \le s} \frac{r_{\{i,j,k\}}}{p_i p_j p_k}$$

The width of this bound is

$$\left| \sum_{1 \le i < j < k \le s} \frac{r_{\{i,j,k\}}}{p_i p_j p_k} \right|.$$

We can tighten the bound further and further by adding more alternating layers until both sides match the exact prime count. The overall complexity of calculating the error is similar to calculating the prime count itself, and this shows why we need a different point of view.

9.5 Dependency Factor

We have defined waves of numbers, and we have also defined the *Independent World*. In that world, the same generators still exist, but their phases are unknown rather than fixed. What the Independent World gives us instead is computability: counts become rational, inclusion–exclusion becomes cleaner, De Morgan's law can be applied, the model moves into multiplicative space, and large structures can be described by exact algebraic products.

Then we need a bridge back to the Dependent World.

Earlier, we defined the total tail error $E_1(n)$, which measures the additive difference between an exact dependent count and its independent model. But when the main structures are multiplicative— products of local factors, densities, and combined spectra—an additive correction is no longer the most natural language. In such settings, it is better to express the discrepancy as a *rate*, that is, as a *factor*.

This is the role of the dependency factor. It converts the gap between the Dependent World and the Independent World into a multiplicative object. In this way, the whole theory remains homogeneous: independent models provide the algebra, and dependency factors provide the correction.

The dependency factor is therefore the main bridge between exact arithmetic structure and its independent approximation.

9.5.1 Dependency Factor Function

For a wave, a spectrum, or any model or pattern that can be defined by the Number–Wave Framework, the dependency factor is the ratio

$$\frac{\text{exact dependent count}}{\text{independent count of the same object}}.$$

For a wave,

$$d\big(w_{p_i}(x, r, s, e)\big) := \frac{\big|w_{p_i}(x, r, s, e)\big|}{\big|w_{p_i}(x, *, s, e)\big|}.$$

For a spectrum S,

$$d\big(S(n)\big) := \frac{|S_{\text{dep}}(n)|}{|S_{\text{ind}}(n)|}.$$

For the Prime Effective Shadow Spectrum,

$$d_{PESS}(n) := d\big(\text{PESS}(x, n)\big) = \frac{|\text{PESS}(x, n)|}{|\text{IPESS}(x, n)|}.$$

Since

$$E_1(n) = |\text{PESS}(x, n)| - |\text{IPESS}(x, n)|,$$

we immediately obtain

$$d_{PESS}(n) = \frac{|\text{IPESS}(x, n)| + E_1(n)}{|\text{IPESS}(x, n)|} = 1 + \frac{E_1(n)}{|\text{IPESS}(x, n)|}.$$

And because

$$|\text{PESS}(x, n)| = \pi(n) - \pi(\sqrt{n}) + 1,$$

and

$$\boxed{|\text{IPESS}(x,n)| = n \prod_{p_i \leq \sqrt{n}} \left(1 - \frac{1}{p_i}\right)}$$

we have

$$\boxed{d_{PESS}(n) = \frac{\pi(n) - \pi(\sqrt{n}) + 1}{n \prod_{p_i \leq \sqrt{n}} \left(1 - \frac{1}{p_i}\right)}.}$$

9.6 Prime Generating Function $G(x,n)$

9.6.1 Definition of $G(x,n)$

The independent spectrum naturally leads to a prime product. We package all independent contributions into a single generating function.

Definition. Fix a real number $x \geq 0$. Define

$$G(x,n) := \prod_{x < p \leq n} \left(1 - \frac{x}{p}\right),$$

where the product runs over primes p.

The cutoff condition $x < p$ is essential. It avoids the zero factor when $p = x$ (if x itself is prime), ensures that every factor is positive, and keeps the product well defined for all real x.

9.6.2 The case $x = 1$ and independent shadows

For $x = 1$,

$$G(1, \sqrt{n}) = \prod_{p \leq \sqrt{n}} \left(1 - \frac{1}{p}\right),$$

which is exactly the independent shadow factor.

Hence the independent decomposition of the prime count can be written as

$$\pi(n) = |\text{PESS}(x,n)| + \pi(\sqrt{n}) - 1,$$

with

$$|\text{PESS}(x,n)| = d_{PESS}(n) \cdot |\text{IPESS}(x,n)| = d_{PESS}(n) \cdot n \cdot G(1, \sqrt{n}).$$

This decomposition separates two components: an independent density governed by $G(1, \sqrt{n})$, and a dependency correction captured by $d(\cdot)$.

9.6.3 Expansion of $G(x,n)$

Let

$$p_1 < p_2 < \cdots < p_s \leq n$$

be the primes up to n.

For the range $0 \leq x < 2$ used below, the cutoff condition $x < p$ excludes no primes, so this notation is consistent. Expanding the product gives

$$G(x, n) = \prod_{i=1}^{s} \left(1 - \frac{x}{p_i} \right)$$

$$= 1 - x \sum_{i=1}^{s} \frac{1}{p_i} + x^2 \sum_{1 \leq i < j \leq s} \frac{1}{p_i p_j} - x^3 \sum_{1 \leq i < j < k \leq s} \frac{1}{p_i p_j p_k} + \cdots .$$

For each $m \geq 0$, define

$$D_m(n) := \sum_{p_{i_1} < \cdots < p_{i_m} \leq n} \frac{1}{p_{i_1} \cdots p_{i_m}}, \qquad D_0(n) = 1.$$

Then

$$G(x, n) = \sum_{m=0}^{\pi(n)} (-1)^m D_m(n) \, x^m.$$

9.6.4 Taylor View

Figure 9.5: Brook Taylor (1685–1731).

Preface on Taylor series.   A Taylor series is another way of seeing a function. In the ordinary view, we study the function across its whole domain.

In the Taylor view, we track a particle moving along the function. If we know its position, its speed, its acceleration, and its higher-order accelerations at a point, then that local information already encodes the whole route to infinity.

Viewed as a function of x,

$$G(0, n) = 1, \qquad G^{(m)}(0, n) = (-1)^m m! \, D_m(n).$$

Equivalently,

$$D_m(n) = \frac{(-1)^m}{m!} \, G^{(m)}(0, n).$$

Thus every multi-prime reciprocal sum is encoded in a single derivative of G. To understand $G(x, n)$ is to understand all independent coefficients simultaneously.

9.6.5   Sum of Reciprocals of Primes

Exact identity

We start from the identity

$$\sum_{p \leq n} \frac{1}{p} = \frac{\pi(n)}{n} + \sum_{i=2}^{n} \frac{\pi(i)}{i(i-1)} - \sum_{p \leq n} \frac{1}{p(p-1)}.$$

Asymptotic behaviour

Using $\pi(x) \sim x/\ln x$,

$$\sum_{i=2}^{n} \frac{\pi(i)}{i(i-1)} \sim \int_{2}^{n} \frac{dx}{x \ln x} = \ln \ln n + O(1).$$

The correction term

$$\sum_{p \leq n} \frac{1}{p(p-1)}$$

converges. Hence the reciprocal-prime sum has the asymptotic form

$$\sum_{p \leq n} \frac{1}{p} = \ln \ln n + B + o(1),$$

for a finite constant B.

For higher powers, the prime reciprocal sums converge:

$$\sum_{p} \frac{1}{p^m} = S_m < \infty, \qquad m \geq 2.$$

9.6.6 Asymptotic Form of $G(x, n)$

Logarithmic expansion

For $0 \le x < 2$,

$$\log G(x, n) = \sum_{p \le n} \log\left(1 - \frac{x}{p}\right).$$

Using

$$\log(1 - u) = -\sum_{m \ge 1} \frac{u^m}{m},$$

we obtain

$$\log G(x, n) = -x \sum_{p \le n} \frac{1}{p} - \sum_{m \ge 2} \frac{x^m}{m} \sum_{p \le n} \frac{1}{p^m}.$$

Substituting the asymptotics,

$$\log G(x, n) = -x \ln \ln n - xB - \sum_{m \ge 2} \frac{x^m}{m} S_m + o(1).$$

Equivalently,

$$\log\big((\ln n)^x G(x, n)\big) = -xB - \sum_{m \ge 2} \frac{x^m}{m} S_m + o(1).$$

Hence

$$(\ln n)^x G(x, n) \longrightarrow \exp\left(-xB - \sum_{m \ge 2} \frac{x^m}{m} S_m\right), \qquad n \to \infty.$$

The special case $x = 1$. For $x = 1$,

$$\log\big((\ln n) G(1, n)\big) = -B - \sum_{m \ge 2} \frac{1}{m} S_m + o(1).$$

By Mertens' third theorem,

$$B + \sum_{m \ge 2} \frac{1}{m} S_m = \gamma,$$

where γ is the Euler–Mascheroni constant. Therefore

$$(\ln n) G(1, n) \longrightarrow e^{-\gamma}, \qquad n \to \infty,$$

and hence

$$G(1, n) \sim \frac{e^{-\gamma}}{\ln n}.$$

The special case $x = 2$. For $x = 2$,

$$G(2, n) = \prod_{2 < p \le n} \left(1 - \frac{2}{p}\right).$$

Finite factorization relative to the $x = 1$ case.   Define

$$C_2(n) := \prod_{2 < p \leq n} \frac{1 - \frac{2}{p}}{\left(1 - \frac{1}{p}\right)^2} = \prod_{2 < p \leq n} \frac{p(p-2)}{(p-1)^2}.$$

Then

$$G(2, n) = \left(\prod_{2 < p \leq n} \left(1 - \frac{1}{p}\right) \right)^2 \cdot C_2(n).$$

Using the $x = 1$ asymptotic,

$$\prod_{p \leq n} \left(1 - \frac{1}{p}\right) \sim \frac{e^{-\gamma}}{\ln n}.$$

Removing the factor $p = 2$ gives

$$\prod_{2 < p \leq n} \left(1 - \frac{1}{p}\right) = \frac{\prod_{p \leq n} \left(1 - \frac{1}{p}\right)}{1 - \frac{1}{2}} \sim 2 \frac{e^{-\gamma}}{\ln n}.$$

Therefore

$$G(2, n) \sim 4 \, e^{-2\gamma} \frac{C_2(n)}{(\ln n)^2}.$$

If the limit exists, we define it separately by

$$C_2 := \lim_{n \to \infty} C_2(n) = \lim_{n \to \infty} \prod_{2 < p \leq n} \frac{p(p-2)}{(p-1)^2}.$$

C_2 is called the Twin Prime Constant.
After taking the limit, the asymptotic becomes

$$G(2, n) \sim 4 \, e^{-2\gamma} \frac{C_2}{(\ln n)^2}, \qquad n \to \infty.$$

The general integer case $x = k$

For an integer $k \geq 1$,

$$G(k, n) = \prod_{k < p \leq n} \left(1 - \frac{k}{p}\right).$$

Define the finite correction factor

$$C_k(n) := \prod_{k < p \leq n} \frac{1 - \frac{k}{p}}{\left(1 - \frac{1}{p}\right)^k} = \prod_{k < p \leq n} \frac{p^{k-1}(p-k)}{(p-1)^k}.$$

Then the exact factorization is

$$G(k, n) = \left(\prod_{k < p \leq n} \left(1 - \frac{1}{p} \right) \right)^k \cdot C_k(n).$$

Using the $x = 1$ case,

$$\prod_{k < p \leq n} \left(1 - \frac{1}{p} \right) = \frac{\prod_{p \leq n} \left(1 - \frac{1}{p} \right)}{\prod_{p \leq k} \left(1 - \frac{1}{p} \right)} \sim \frac{e^{-\gamma}}{\ln n} \cdot \frac{1}{\prod_{p \leq k} \left(1 - \frac{1}{p} \right)}.$$

Therefore

$$G(k, n) \sim \frac{e^{-k\gamma}}{(\ln n)^k} \frac{C_k(n)}{\left(\prod_{p \leq k} \left(1 - \frac{1}{p} \right) \right)^k}.$$

If the limit exists, define it separately by

$$C_k := \lim_{n \to \infty} C_k(n) = \lim_{n \to \infty} \prod_{k < p \leq n} \frac{p^{k-1}(p - k)}{(p - 1)^k}.$$

Then

$$G(k, n) \sim \frac{e^{-k\gamma}}{(\ln n)^k} \frac{C_k}{\left(\prod_{p \leq k} \left(1 - \frac{1}{p} \right) \right)^k}, \qquad n \to \infty.$$

Figure 9.6: Franz Mertens (1840–1927).

Summary

At this stage, all structural dependency has been isolated.

Invoking the Prime Number Theorem together with Mertens' theorem yields the asymptotic limit

$$d_{PESS}(n) \longrightarrow \frac{e^{\gamma}}{2}.$$

Hence

$$\pi(n) = d_{PESS}(n) \cdot |\text{IPESS}(x, n)| + \pi(\sqrt{n}) - 1.$$

Asymptotically,

$$\pi(n) \sim \left(\frac{e^{\gamma}}{2} \right) n \, G(1, \sqrt{n}).$$

Consider the universe

$$U_n = [1, n],$$

with effective primes

$$p_1 < p_2 < \cdots < p_s \leq \sqrt{n}.$$

For each effective prime, we use the residue space

$$R_{p_i} = \{0, 1, \ldots, p_i - 1\}.$$

We also use the primorial of the last effective prime:

$$p_s\# = \prod_{i=1}^{s} p_i.$$

Let $p_\ell \leq n$ denote the last prime before n.

9.6.7 Prime Effective Spectrum (PES)

The Prime Effective Spectrum is the set of points crossed by the effective prime waves on $[1, n]$:

$$\mathrm{PES}(x, n) = \bigcup_{i=1}^{s} w_{p_i}(x, 0, 1, n).$$

Its size is

$$\left| \mathrm{PES}(x, n) \right| = n - \left(\pi(n) - \pi(\sqrt{n}) + 1 \right).$$

9.6.8 Prime Effective Shadow Spectrum (PESS)

The Prime Effective Shadow Spectrum is the shadow of the Prime Effective Spectrum:

$$\mathrm{PESS}(x, n) = \bigcap_{i=1}^{s} w_{p_i}(x, \, R_{p_i} \setminus \{0\}, \, 1, \, n).$$

Its size is

$$\left| \mathrm{PESS}(x, n) \right| = \pi(n) - \pi(\sqrt{n}) + 1.$$

9.6.9 Independent Prime Effective Shadow Spectrum (IPESS)

The Independent Prime Effective Shadow Spectrum is the independent analogue of $\mathrm{PESS}(x, n)$:

$$\mathrm{IPESS}(x, n) = \bigcap_{i=1}^{s} \neg\, w_{p_i}(x, *, 1, n).$$

The dependency factor links the dependent and independent models:

$$\left| \mathrm{PESS}(x, n) \right| = d_{PESS}(n) \cdot \left| \mathrm{IPESS}(x, n) \right|.$$

And at infinity:

$$\lim_{n \to \infty} d_{PESS}(n) = \frac{1}{2e^{-\gamma}} = \frac{e^{\gamma}}{2}.$$

Also, the independent size is given by the prime product function:

$$\left| \mathrm{IPESS}(x, n) \right| = n \cdot G(1, \sqrt{n}).$$

We define the prime product function by

$$G(1, \sqrt{n}) = \prod_{i=1}^{s} \left(1 - \frac{1}{p_i} \right).$$

9.7   Prime Shadow Spectrum (PSS)

9.8   Prime Spectra

The Prime Shadow Spectrum is equivalent to the distribution of composite numbers:

$$\text{PSS}(x, n) = \bigcup_{i=1}^{s} w_{p_i}(x, 0, p_i^2, n).$$

In this section, the start point of each shadow wave is p_i^2.

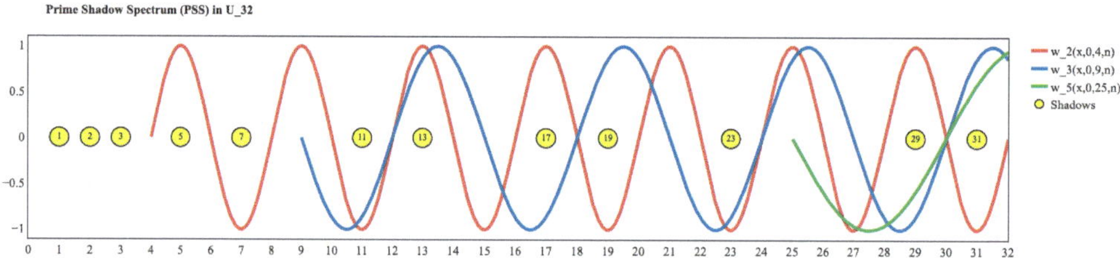

Figure 9.7: $\text{PSS}(x, 32)$.

$$|\text{PSS}(x, n)| = n - (\pi(n) + 1).$$

9.9   Prime Spectrum (PS) - Distribution of Primes

The distribution of prime numbers can be defined by the Prime Spectrum:

$$\text{PS}(x, n) = \bigcap_{i=1}^{s} \neg\, w_{p_i}(x, 0, p_i^2, n) = \bigcap_{i=1}^{s} w_{p_i}(x, R_{p_i} \setminus \{0\}, p_i^2, n).$$

The Prime Spectrum shows all prime numbers, together with 1.

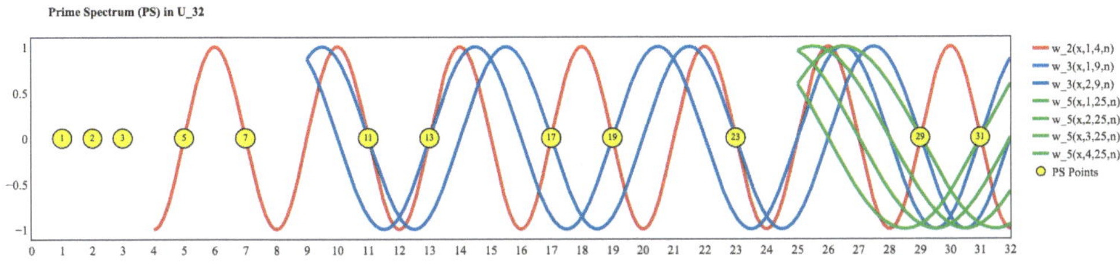

Figure 9.8: $\mathrm{PS}(x, 32)$.

$$|\mathrm{PS}(x, n)| = \pi(n) + 1.$$

Connection between PS, PSS, PES, and PESS

The four spectra are related transparently as follows:

$$\begin{aligned}
\mathrm{PSS}(x, n) &= \{\text{composites in } [1, n]\}, \\
\mathrm{PESS}(x, n) &= \{1\} \cup \{\text{primes in } (\sqrt{n}, n]\}, \\
\mathrm{PES}(x, n) &= \{\text{effective primes}\} \cup \{\text{composites in } [1, n]\}, \\
\mathrm{PS}(x, n) &= \{1\} \cup \{\text{effective primes}\} \cup \{\text{primes in } (\sqrt{n}, n]\}.
\end{aligned}$$

The effective primes are exactly

$$p_1 < p_2 < \cdots < p_s \leq \sqrt{n}.$$

The Prime Shadow Spectrum $\mathrm{PSS}(x, n)$ contains all composite numbers in $[1, n]$, while the Prime Spectrum $\mathrm{PS}(x, \sqrt{n})$ contains

$$\{1\} \cup \{p_1, p_2, \ldots, p_s\}.$$

Therefore,

$$\mathrm{PES}(x, n) = \mathrm{PSS}(x, n) \cup \big(\mathrm{PS}(x, \sqrt{n}) \setminus \{1\}\big).$$

Equivalently,

$$\mathrm{PES}(x, n) = \big(\mathrm{PSS}(x, n) \cup \neg\mathrm{PSS}(x, \sqrt{n})\big) \setminus \{1\},$$

where the complement $\neg\mathrm{PSS}(x, \sqrt{n})$ is taken inside the smaller universe $[1, \sqrt{n}]$.

Taking cardinalities gives

$$|\mathrm{PES}(x, n)| = |\mathrm{PSS}(x, n)| + \pi(\sqrt{n}).$$

Also,

$$|\mathrm{PS}(x, \sqrt{n})| = \sqrt{n} - |\mathrm{PSS}(x, \sqrt{n})|,$$

so

$$\pi(\sqrt{n}) = \sqrt{n} - |\mathrm{PSS}(x, \sqrt{n})| - 1.$$

Hence,

$$|\mathrm{PES}(x, n)| = |\mathrm{PSS}(x, n)| + \sqrt{n} - |\mathrm{PSS}(x, \sqrt{n})| - 1.$$

Now split the full prime spectrum at the level $\sqrt{n}$. The points of $\mathrm{PESS}(x, n)$ are exactly

$$\{1\} \cup \{p \leq n : \ p \text{ is prime and } p > \sqrt{n}\}.$$

Hence

$$\mathrm{PS}(x, n) = \mathrm{PESS}(x, n) \cup \left(\mathrm{PS}(x, \sqrt{n}) \setminus \{1\}\right),$$

and therefore

$$|\mathrm{PS}(x, n)| = |\mathrm{PESS}(x, n)| + \pi(\sqrt{n}).$$

Equivalently,

$$\mathrm{PESS}(x, n) = \mathrm{PS}(x, n) \setminus \left(\mathrm{PS}(x, \sqrt{n}) \setminus \{1\}\right).$$

9.9.1 Independent Prime Spectrum (IPS)

The Independent Prime Spectrum is the independent version of $\mathrm{PS}(x, n)$: all deterministic phases are replaced by an unknown phase on the same domain.

$$\mathrm{IPS}(x, n) = \bigcap_{i=1}^{s} \neg\, w_{p_i}(x, *, p_i^2, n) = \neg \bigcup_{i=1}^{s} w_{p_i}(x, *_{-1}, p_i^2, n).$$

Since each prime wave loses exactly one phase, the survival factor contributed by p_i is

$$1 - \frac{1}{p_i},$$

and we denote this one-phase loss by $*_{-1}$.

Now we calculate the value of the Independent Prime Spectrum.

The formula is computed by prime-square blocks:

$$|\mathrm{IPS}(x, n)| = 3 + \sum_{i=2}^{s} (p_i^2 - p_{i-1}^2) \prod_{j=1}^{i-1} \left(1 - \frac{1}{p_j}\right) + (n - p_s^2) \prod_{j=1}^{s} \left(1 - \frac{1}{p_j}\right).$$

Let us go through an example.

Example ($n = 200$). Using the actual prime-square blocks up to 200, we obtain

$$[1, 4) : \quad (4 - 1) \cdot 1 = 3$$

$$[4, 9) : \quad (9 - 4) \cdot \tfrac{1}{2} = 2.50$$

$$[9, 25) : \quad (25 - 9) \cdot \tfrac{1}{2} \cdot \tfrac{2}{3} = 5.33$$

$$[25, 49) : \quad (49 - 25) \cdot \tfrac{1}{2} \cdot \tfrac{2}{3} \cdot \tfrac{4}{5} = 6.40$$

$$[49, 121) : \quad (121 - 49) \cdot \tfrac{1}{2} \cdot \tfrac{2}{3} \cdot \tfrac{4}{5} \cdot \tfrac{6}{7} = 16.46$$

$$[121, 169) : \quad (169 - 121) \cdot \tfrac{1}{2} \cdot \tfrac{2}{3} \cdot \tfrac{4}{5} \cdot \tfrac{6}{7} \cdot \tfrac{10}{11} = 9.97$$

$$[169, 200] : \quad (200 - 169) \cdot \tfrac{1}{2} \cdot \tfrac{2}{3} \cdot \tfrac{4}{5} \cdot \tfrac{6}{7} \cdot \tfrac{10}{11} \cdot \tfrac{12}{13} = 5.95$$

Summing these block contributions gives

$$|\mathrm{IPS}(x, 200)| = 49.61 + \epsilon,$$

where ϵ absorbs the remaining decimal-rounding discrepancy.

To rewrite the block sum in integral form, recall

$$G(x, n) := \prod_{x < p \leq n} \left(1 - \frac{x}{p}\right).$$

In the special case $x = 1$, this becomes

$$G(1, u) = \prod_{p \leq u} \left(1 - \frac{1}{p}\right).$$

Let

$$p_s \leq \sqrt{n} < p_{s+1}.$$

Then the exact finite sum for $|\mathrm{IPS}(x, n)|$ is

$$\boxed{|\mathrm{IPS}(x, n)| = 3 + \sum_{i=2}^{s} \left(p_i^2 - p_{i-1}^2\right) G(1, p_{i-1}) + \left(n - p_s^2\right) G(1, p_s).}$$

This is an exact rewriting of the prime-square block sum.

Indeed, on each interval

$$p_{i-1} \leq u < p_i,$$

no new prime appears, so $G(1, u)$ remains constant:

$$G(1, u) = G(1, p_{i-1}).$$

Therefore each block contribution can be written exactly as

$$\left(p_i^2 - p_{i-1}^2\right) G(1, p_{i-1}) = \int_{p_{i-1}}^{p_i} 2u\, G(1, u)\, du,$$

using the change of variable

$$t = u^2, \qquad dt = 2u\, du.$$

Summing over all full blocks, and treating the final partial block in the same way, gives

$$\boxed{|\mathrm{IPS}(x, n)| = 3 + 2 \int_{2}^{\sqrt{n}} u\, G(1, u)\, du.}$$

Since $G(1, u) = 1$ on the interval $1 \leq u < 2$, the initial term 3 can also be absorbed into the integral:

$$3 = \int_{1}^{2} 2u\, du.$$

Hence we obtain the exact integral form

$$\boxed{|\mathrm{IPS}(x, n)| = 2 \int_{1}^{\sqrt{n}} u\, G(1, u)\, du.}$$

So

$$\left| \mathrm{IPS}(x,n) \right| = \left| \bigcap_{i=1}^{s} \neg\, w_{p_i}(x, *, p_i^2, n) \right| = \left| \neg \bigcup_{i=1}^{s} w_{p_i}(x, *_{-1}, p_i^2, n) \right|$$

$$= 2 \int_{1}^{\sqrt{n}} u\, G(1,u)\, du.$$

Using the asymptotic formula

$$G(1,u) \sim \frac{e^{-\gamma}}{\ln u},$$

the asymptotic form becomes

$$\left| \mathrm{IPS}(x,n) \right| \sim 3 + 2e^{-\gamma} \int_{2}^{\sqrt{n}} \frac{u}{\ln u}\, du.$$

After the substitution $t = u^2$, this becomes

$$\left| \mathrm{IPS}(x,n) \right| \sim 3 + 2e^{-\gamma} \int_{4}^{n} \frac{dt}{\ln t}.$$

Equivalently,

$$\left| \mathrm{IPS}(x,n) \right| \sim 2e^{-\gamma}\, \mathrm{li}(n) = d_{PS}\, \mathrm{li}(n),$$

where d_{PS} is the prime dependency factor at infinity.

9.10   Prime Counting Function

9.10.1   Prime Counting Function (Exact)

In the previous chapter, we obtained

$$\pi(n) = \left| \mathrm{PESS}(x,n) \right| + \pi(\sqrt{n}) - 1 = n - \left| \mathrm{PES}(x,n) \right| + \pi(\sqrt{n}) - 1.$$

Since $\left| \mathrm{PES}(x,n) \right|$ can be computed exactly by inclusion–exclusion, this recursion gives an exact computation of $\pi(n)$.

There is also a more direct form based on the Prime Shadow Spectrum:

$$\pi(n) = \left| \mathrm{PS}(x,n) \right| - 1 = n - \left| \mathrm{PSS}(x,n) \right| - 1.$$

Again, $\left| \mathrm{PSS}(x,n) \right|$ can be computed exactly by inclusion–exclusion, so this formula also gives the exact value of the prime-counting function.

Example: exact computation of $\pi(32)$ using PSS.   For $n = 32$,

$$\pi(32) = 32 - \left| \mathrm{PSS}(x,32) \right| - 1.$$

So we first compute $\left| \mathrm{PSS}(x,32) \right|$ exactly.

Since the effective primes are $2, 3, 5$, we have

$$\mathrm{PSS}(x,32) = w_2(x,0,4,32) \cup w_3(x,0,9,32) \cup w_5(x,0,25,32).$$

We use the wave-counting formula

$$\left| w_T(x, r, s, e) \right| = \left\lfloor \frac{T + e - \left(s + ((r - s) \bmod T)\right)}{T} \right\rfloor.$$

By inclusion–exclusion,

$$\begin{aligned}
\left| \mathbf{PSS}(x, 32) \right| = &\ |w_2(x, 0, 4, 32)| + |w_3(x, 0, 9, 32)| + |w_5(x, 0, 25, 32)| \\
&- |w_6(x, 0, 9, 32)| - |w_{10}(x, 0, 25, 32)| - |w_{15}(x, 0, 25, 32)| \\
&+ |w_{30}(x, 0, 25, 32)|.
\end{aligned}$$

The intersection waves are

$$w_2(x, 0, 4, 32) \cap w_3(x, 0, 9, 32) = w_6(x, 0, 9, 32),$$

$$w_2(x, 0, 4, 32) \cap w_5(x, 0, 25, 32) = w_{10}(x, 0, 25, 32),$$

$$w_3(x, 0, 9, 32) \cap w_5(x, 0, 25, 32) = w_{15}(x, 0, 25, 32),$$

$$w_2(x, 0, 4, 32) \cap w_3(x, 0, 9, 32) \cap w_5(x, 0, 25, 32) = w_{30}(x, 0, 25, 32).$$

Now compute each term:

$$|w_2(x, 0, 4, 32)| = \left\lfloor \frac{2 + 32 - \left(4 + ((0 - 4) \bmod 2)\right)}{2} \right\rfloor = \left\lfloor \frac{30}{2} \right\rfloor = 15,$$

$$|w_3(x, 0, 9, 32)| = \left\lfloor \frac{3 + 32 - \left(9 + ((0 - 9) \bmod 3)\right)}{3} \right\rfloor = \left\lfloor \frac{26}{3} \right\rfloor = 8,$$

$$|w_5(x, 0, 25, 32)| = \left\lfloor \frac{5 + 32 - \left(25 + ((0 - 25) \bmod 5)\right)}{5} \right\rfloor = \left\lfloor \frac{12}{5} \right\rfloor = 2,$$

$$|w_6(x, 0, 9, 32)| = \left\lfloor \frac{6 + 32 - \left(9 + ((0 - 9) \bmod 6)\right)}{6} \right\rfloor = \left\lfloor \frac{26}{6} \right\rfloor = 4,$$

$$|w_{10}(x, 0, 25, 32)| = \left\lfloor \frac{10 + 32 - \left(25 + ((0 - 25) \bmod 10)\right)}{10} \right\rfloor = \left\lfloor \frac{12}{10} \right\rfloor = 1,$$

$$|w_{15}(x, 0, 25, 32)| = \left\lfloor \frac{15 + 32 - \left(25 + ((0 - 25) \bmod 15)\right)}{15} \right\rfloor = \left\lfloor \frac{17}{15} \right\rfloor = 1,$$

$$|w_{30}(x, 0, 25, 32)| = \left\lfloor \frac{30 + 32 - \left(25 + ((0 - 25) \bmod 30)\right)}{30} \right\rfloor = \left\lfloor \frac{32}{30} \right\rfloor = 1.$$

Therefore,

$$\left| \mathbf{PSS}(x, 32) \right| = 15 + 8 + 2 - 4 - 1 - 1 + 1 = 20.$$

Hence

$$\pi(32) = 32 - 20 - 1 = 11.$$

Its complexity is

$$O\left(2^{\pi(\sqrt{n})} \right).$$

9.10.2　Direct Prime Counting Function Formula

The prime counting function $\pi(n)$ can be written exactly in the following inclusion–exclusion form:

$$\pi(n) = \sum_{p \leq n} \left\lfloor \frac{p^2 - 1}{p} \right\rfloor - \sum_{p < q \leq n} \left\lfloor \frac{q^2 - 1}{pq} \right\rfloor + \sum_{p < q < r \leq n} \left\lfloor \frac{r^2 - 1}{pqr} \right\rfloor$$
$$- \sum_{p < q < r < s \leq n} \left\lfloor \frac{s^2 - 1}{pqrs} \right\rfloor + \sum_{p < q < r < s < t \leq n} \left\lfloor \frac{t^2 - 1}{pqrst} \right\rfloor - \cdots$$

where all sums are taken over increasing prime tuples $\leq n$, and in each term the numerator is determined by the largest prime in that tuple.

This is an exact formula for the prime counting function. It is not intended for practical computation, since it involves all nonempty subsets of the primes up to n. Hence its complexity is

$$O\left(2^{\pi(n)}\right),$$

so the number of terms grows exponentially with $\pi(n)$.

In fact, the choice $p^2 - 1$ in the numerator is not unique. More generally, if for each prime m we choose any integer a_m satisfying

$$m \leq a_m \leq m^2 - 1,$$

then the same inclusion–exclusion structure still gives exactly $\pi(n)$. Thus we have the more general identity

$$\pi(n) = \sum_{p \leq n} \left\lfloor \frac{a_p}{p} \right\rfloor - \sum_{p < q \leq n} \left\lfloor \frac{a_q}{pq} \right\rfloor + \sum_{p < q < r \leq n} \left\lfloor \frac{a_r}{pqr} \right\rfloor$$
$$- \sum_{p < q < r < s \leq n} \left\lfloor \frac{a_s}{pqrs} \right\rfloor + \sum_{p < q < r < s < t \leq n} \left\lfloor \frac{a_t}{pqrst} \right\rfloor - \cdots$$

for any choice of integers a_m with

$$m \leq a_m \leq m^2 - 1.$$

9.11　Dependency Factor Principle

The dependency factor absorbs all dependencies of a dependent model into a single value, so that the output corresponds to the independent model. When multiplied into the independent model, it returns the value of the dependent model.

Dependency Factor Principle.

If $d_A(n) = d_A\big(1 + O(\varepsilon_A(n))\big)$ and $d_B(n) = d_B\big(1 + O(\varepsilon_B(n))\big)$,

then $d_{A \cap B}(n) = d_A d_B\big(1 + O(\varepsilon_A(n) + \varepsilon_B(n) + \varepsilon_A(n)\varepsilon_B(n))\big)$.

We consider four types of dependency-factor behaviour.
Consider two spectra A and B.

A. Fixed dependency factors   The ideal case is that the dependency factor is a fixed number for all n. So, if they have fixed dependency factors,

$$d_A(n) = d_A, \qquad d_B(n) = d_B,$$

then

$$|A_{\text{dep}}(x,n)| = d_A\,|A_{\text{ind}}(x,n)|, \qquad |B_{\text{dep}}(x,n)| = d_B\,|B_{\text{ind}}(x,n)|.$$

Then

$$|A_{\text{dep}}(x,n) \cap B_{\text{dep}}(x,n)| = d_A\,d_B\,|A_{\text{ind}}(x,n) \cap B_{\text{ind}}(x,n)|.$$

We can also extend this to other set operations:

$$\boxed{\begin{aligned}|A_{\text{dep}}(n) \cup B_{\text{dep}}(n)| &= d_A\,|A_{\text{ind}}(n)| + d_B\,|B_{\text{ind}}(n)| \\ &\quad - d_A\,d_B\,|A_{\text{ind}}(n) \cap B_{\text{ind}}(n)|.\end{aligned}}$$

Generally, if the dependency factor is a fixed value, then d_A and d_B must be either one or zero. The ideal case can happen only for an empty spectrum or a full spectrum. There is no other way for the dependency factor to remain constant on the whole domain.

Now consider one of the simplest spectra, consisting only of $w_2(x,0)$. Its dependency factor is

$$d_{w_2}(n) = \frac{\lfloor \frac{n}{2} \rfloor}{\frac{n}{2}} = 1 + O(1/n).$$

This shows that the dependency-factor function always carries an error. If the error vanishes at infinity, then the principle becomes exact at infinity. But when we work on finite numbers, we must control the error.

B. Fixed-wave spectra   Examples of fixed-wave spectra: Prime Filter Spectrum (PFS), Prime Candidates Spectrum (PCS).

If the number of waves in the spectrum is fixed while n grows, then we should expect an error at most on the scale of one period, just as in the example of $d_{w_2}(n)$. In this case, the period could be very large, but in any case it is still fixed, and compared to infinity it is nothing.

So

$$d_A(n) = d_A + O(1/n), \qquad d_B(n) = d_B + O(1/n).$$

$$|A_{\text{dep}}(x,n)| = \left(d_A + O(1/n)\right)|A_{\text{ind}}(x,n)|, \qquad |B_{\text{dep}}(x,n)| = \left(d_B + O(1/n)\right)|B_{\text{ind}}(x,n)|.$$

Then we should expect

$$|A_{\text{dep}}(x,n) \cap B_{\text{dep}}(x,n)| = d_A\,d_B\left(1 + O(1/n) + O(1/n^2)\right)|A_{\text{ind}}(x,n) \cap B_{\text{ind}}(x,n)|.$$

C. Effective prime spectra   Examples of effective prime spectra: Prime Spectrum (PS), Prime Shadow Spectrum (PSS), Prime Effective Spectrum (PES), Prime Effective Shadow Spectrum (PESS).

Effective prime waves are waves generated by prime generators $p_i \le \sqrt{n}$. So as n grows, new lower-frequency waves are incrementally added to the model.

If the spectra are defined by effective prime waves, then we can show

$$d_A(n) = d_A + O(1/\ln n), \qquad d_B(n) = d_B + O(1/\ln n).$$

$$|A_{\mathrm{dep}}(x,n)| = \bigl(d_A + O(1/\ln n)\bigr)\,|A_{\mathrm{ind}}(x,n)|, \qquad |B_{\mathrm{dep}}(x,n)| = \bigl(d_B + O(1/\ln n)\bigr)\,|B_{\mathrm{ind}}(x,n)|.$$

And by the Dependency Factor Principle we should have

$$|A_{\mathrm{dep}}(x,n) \cap B_{\mathrm{dep}}(x,n)| = d_A\,d_B\,\bigl(1 + O(1/\ln n) + O(1/\ln^2 n)\bigr)\,|A_{\mathrm{ind}}(x,n) \cap B_{\mathrm{ind}}(x,n)|.$$

Specifically, we can prove that

$$d_{PS}(n) := \frac{|\mathrm{PS}(x,n)|}{|\mathrm{IPS}(x,n)|}.$$

We obtained IPS as a continuous function:

$$\boxed{|\mathrm{IPS}(x,n)| \sim 2e^{-\gamma}\,\mathrm{li}(n).}$$

And

$$\boxed{|\mathrm{PS}(x,n)| \sim \mathrm{li}(n).}$$

Therefore,

$$\boxed{d_{\mathrm{PS}}(n) = \frac{e^{\gamma}}{2}\left(1 + O\!\left(\frac{1}{\ln n}\right)\right).}$$

Similarly,

$$\boxed{d_{\mathrm{PESS}}(n) = \frac{e^{\gamma}}{2}\left(1 + O\!\left(\frac{1}{\ln n}\right)\right).}$$

D. Nonlinear dependency factor   In our context, the dependency factor should be of the form

$$d + o(1).$$

If not, then the relationship between the dependent model and the independent equivalent model is not asymptotically linear, and such models are not part of our present context. So before applying the Dependency Factor Principle, we must always check the order of the error in order to control it.

The Dependency Factor Principle is the key to unlocking other prime problems, and the Prime Dependency Factor is one of the most important constants in mathematics.

It is tuned at infinity, and by algebra over spectra, we can begin to unlock other prime patterns at infinity, also with controlled error on finite numbers.

Chapter 10

Twin Primes

"It will be another million years, at least, before we understand the primes." — *Erdős*

Figure 10.1: Paul Erdős (1913–1996).

10.1   Preface

This is the final chapter of Season One of *Numbers Discoveries*, and its target is twin primes.

In this chapter, we uncover the distribution of twin primes. We derive the exact twin-prime counting function. We reach their value at infinity. We reveal their relation to the prime counting function.

We dedicate this chapter to G. H. Hardy and Paul Erdős—Hardy for helping shape the modern vision of twin primes, and for bringing unmatched clarity, rigor, and elegance to analytic number theory; and Erdős for proving, with Selberg, that even the Prime Number Theorem could be reached by elementary means, while turning mathematics itself into a living world of problems, ideas, and collaboration.

G. H. Hardy: The Voice Behind Great Conjectures

G. H. Hardy gave number theory a rare kind of strength: exact reasoning joined with intellectual style. In the story of twin primes, his influence is monumental. Through his work with Littlewood, he helped shape the modern language and vision of prime patterns, giving mathematicians one of the boldest and most influential pictures of how twin primes should behave. Hardy did not merely advance the subject—he gave it standards. He showed that even the wildest questions in number theory could be approached with precision, depth, and beauty. His legacy lives not only in results, but in the way mathematicians learned to think.

Paul Erdős: The Wandering Fire of Problems

Paul Erdős lived like a force of nature: traveling endlessly, collaborating freely, and scattering deep problems across mathematics like sparks. He helped prove, with Selberg, that the Prime Number Theorem could be established by elementary methods—a stunning achievement that showed how far pure ingenuity could go without relying on heavier machinery. But Erdős is remembered for even more than great proofs. He turned mathematics into a global conversation, where insight mattered more than position, and where a single beautiful problem could inspire generations. In his hands, mathematics was not only a discipline—it was a living adventure.

10.2   Twin Primes

Definition.   Twin primes are pairs of the form $(n - 2, n)$ such that both $n - 2$ and n are prime.

Convention.   We call n a twin prime if both $n - 2$ and n are prime.

Note.   Our twin-prime definition differs from the usual textbook definition, which uses $(n, n + 2)$. In our framework, the form $(n - 2, n)$ is more natural, because it keeps the function $\pi_2(n)$ dependent on the present and the past, rather than on future values such as $\pi(n + 2)$.

Now let us formulate the spectrum that shows all numbers that are not twin primes. Once that spectrum becomes visible, the location of twin primes becomes visible as well.

10.2.1   Twin Prime Shadow Spectrum (TPSS)

Formulating a spectrum is straightforward: we translate the problem into the language of waves. We can define the wave structure directly from the definition, or we can use algebra on known spectra to reach the result more efficiently. In this chapter, we present different approaches in order to provide a deeper conceptual understanding.

In the chapter *Prime Numbers*, we formulated the Prime Shadow Spectrum PSS, in this chapter we need to shift spectrums so lets expand our notation to support shift, with k.

$$\boxed{\mathrm{PSS}(x, k, p_i^2, n) = \bigcup_{i=1}^{s} w_{p_i}(x, k, p_i^2, n)}$$

We also may not write when $k = 0$

$$\mathrm{PSS}(x, n) = \mathrm{PSS}^0(x, n)$$

we can write

$$\mathrm{TPSS}(x, n) = \mathrm{PSS}^0(x, n) \cup \mathrm{PSS}(x, 2, p_i^2, n)$$

Therefore

$$\mathrm{TPSS}(x, n) = \bigcup_{p_i \leq \sqrt{n}} \left\{ w_{p_i}(x, 0, p_i^2, n),\ w_{p_i}(x, 2, p_i^2, n) \right\}.$$

Then we combine the phases and write

$$\boxed{\mathrm{TPSS}(x, n) = \bigcup_{i=1}^{s} w_{p_i}(x, \{0, 2\}, p_i^2, n)}$$

This is the Twin Prime Shadow Spectrum. Its shadow consists of twin primes, excluding the 3 non-twin numbers before the starting point of the first prime wave w_2, which starts at 4.

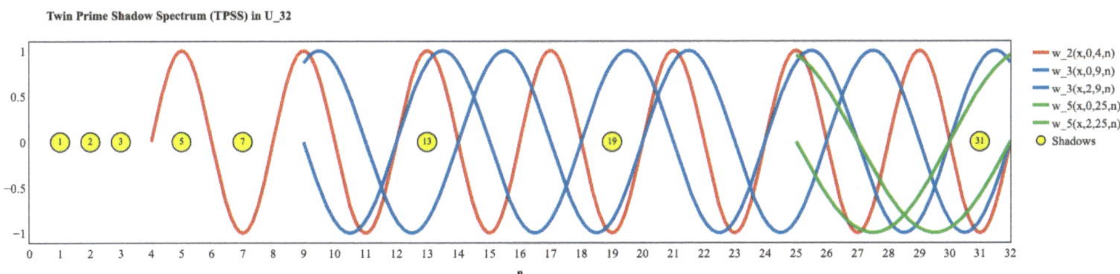

Figure 10.2: TPSS(x,32)

So, except for $\{1, 2, 3\}$, all shadow points of TPSS are twin primes.

We have now formulated TPSS as a union of waves, so the exact twin-prime counting function is available. In fact, the counting functions of all prime patterns are available: we only need to write their spectrum, and for twin primes, as you see, this is easy.

10.2.2 Twin Prime Counting Function — Exact

Now the route to calculate the exact twin-prime counting function is open. Starting from $\mathrm{TPSS}(x, n)$, we see that, except for $\{1, 2, 3\}$, all shadows of TPSS are twin primes. Therefore

$$\boxed{\pi_2(n) = n - |\mathrm{TPSS}(x, n)| - 3}$$

and we have already seen that

$$\mathrm{TPSS}(x, n) = \bigcup_{p_i \leq \sqrt{n}} \left(w_{p_i}(x, 0, p_i^2, n) \cup w_{p_i}(x, 2, p_i^2, n) \right)$$

The count of this union can be calculated using inclusion–exclusion, wave algebra, and the wave-counting function. Let us go through an example and count the exact number of twin primes below 24.

Example: $\pi_2(24)$.   Here $\sqrt{24} < 5$, so the effective primes are $\{2, 3\}$. Also $w_2(x, 2) = w_2(x, 2 \bmod 2) = w_2(x, 0)$, hence

$$\mathrm{TPSS}(x, 24) = w_2(x, 0, 4, 24) \cup w_3(x, 0, 9, 24) \cup w_3(x, 2, 9, 24)$$
$$= \{w_2(x, 0, 4, 24), w_3(x, 0, 9, 24), w_3(x, 2, 9, 24)\}.$$

Parallel waves are fundamental dependency. They must be considered on all models. Here the parallel waves $w_3(x, 0, 9, 24)$ and $w_3(x, 2, 9, 24)$ are disjoint, so all the terms that have both of them never shape a focal point, so they get disjoint.

$$|w_3(x, 0, 9, 24) \cap w_3(x, 2, 9, 24)| = 0.$$

By the wave intersection rules,

$$w_2(x, 0) \cap w_3(x, 0) = w_6(x, 0), \qquad w_2(x, 0) \cap w_3(x, 2) = w_6(x, 2),$$

with effective domain $[3^2, 24] = [9, 24]$:

$$w_2(x, 0, 4, 24) \cap w_3(x, 0, 9, 24) = w_6(x, 0, 9, 24),$$

$$w_2(x, 0, 4, 24) \cap w_3(x, 2, 9, 24) = w_6(x, 2, 9, 24).$$

So inclusion–exclusion reduces to

$$|\mathrm{TPSS}(x, 24)| = |w_2(x, 0, 4, 24)| + |w_3(x, 0, 9, 24)| + |w_3(x, 2, 9, 24)|$$
$$- |w_6(x, 0, 9, 24)| - |w_6(x, 2, 9, 24)|.$$

Now compute the wave-counting function:

$$|w_2(x, 0, 4, 24)| = \left\lfloor \tfrac{24+2-0}{2} \right\rfloor - \left\lfloor \tfrac{(2^2-1)+2-0}{2} \right\rfloor = 13 - 2 = 11,$$

$$|w_3(x, 0, 9, 24)| = \left\lfloor \tfrac{24+3-0}{3} \right\rfloor - \left\lfloor \tfrac{(3^2-1)+3-0}{3} \right\rfloor = 9 - 3 = 6,$$

$$|w_3(x, 2, 9, 24)| = \left\lfloor \tfrac{24+3-2}{3} \right\rfloor - \left\lfloor \tfrac{(3^2-1)+3-2}{3} \right\rfloor = 8 - 3 = 5,$$

$$|w_6(x, 0, 9, 24)| = \left\lfloor \tfrac{24+6-0}{6} \right\rfloor - \left\lfloor \tfrac{(3^2-1)+6-0}{6} \right\rfloor = 5 - 2 = 3,$$

$$|w_6(x, 2, 9, 24)| = \left\lfloor \tfrac{24+6-2}{6} \right\rfloor - \left\lfloor \tfrac{(3^2-1)+6-2}{6} \right\rfloor = 4 - 2 = 2.$$

Thus

$$|\mathrm{TPSS}(x, 24)| = (11 + 6 + 5) - (3 + 2) = 17,$$

and therefore

$$\pi_2(24) = 24 - 17 - 3 = 4.$$

This is the exact twin-prime counting function. However, the number of terms grows exponentially like

$$\sim 2^{(2\cdot\pi(\sqrt{n})-1)}.$$

So this method is not suitable for large n. In the following sections, we seek a better formulation that works for large numbers and shows us what happens at infinity.

But before that, let us now show the spectrum of twin primes, a formula that shows the distribution of twin primes.

10.2.3   Twin Prime Spectrum (TPS)

We have different ways to reach our goal result. We can use operation on known spectrums or we can directly equip waves based on the definition. The twin-prime condition means that both phase 0 and phase 2 must be avoided. So simply the definition tells

$$\text{TPS}(x,n) = \bigcap_{i=1}^{s}\left(\neg w_{p_i}(x,0,p_i^2,n) \,\cap\, \neg w_{p_i}(x,2,p_i^2,n)\right)$$

We project it by two negations and transfer them into the problem and apply De Morgan, so we have

$$\text{TPS}(x,n) = \bigcap_{i=1}^{s} \neg\left(w_{p_i}(x,0,p_i^2,n) \,\cup\, w_{p_i}(x,2,p_i^2,n)\right)$$

So we reach a satisfying formulation of twin-prime distribution:

$$\text{TPS}(x,n) = \bigcap_{i=1}^{s} w_{p_i}(x,\, R_{p_i} \setminus \{0,2\},\, p_i^2,\, n)$$

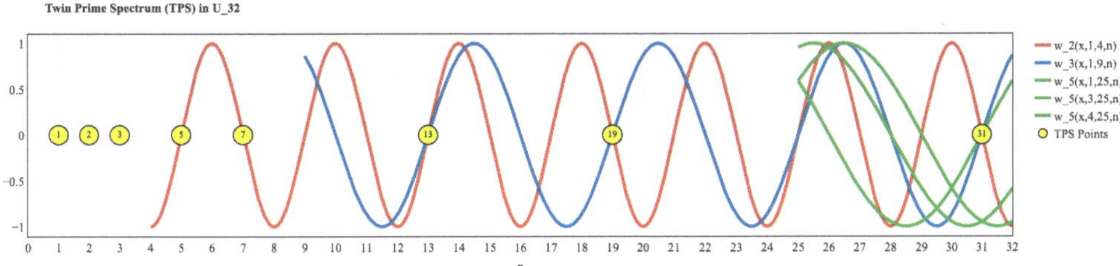

Figure 10.3: TPS(x,32)

The distribution of twin primes is equivalent to the twin-prime spectrum, excluding the points $\{1,2,3\}$.

We also have a better way to reach $\text{TPS}(x,n)$. It may be simpler to define the Twin Prime Spectrum directly from the Twin Prime Shadow Spectrum:

$$\text{TPS}(x,n) = \neg\text{TPSS}(x,n) = \neg\bigcup_{i=1}^{s} w_{p_i}(x,\{0,2\},p_i^2,n) = \bigcap_{i=1}^{s} w_{p_i}(x,\, R_{p_i} \setminus \{0,2\},\, p_i^2,\, n)$$

Or even make it by Prime Spectrum PS.

10.2.4   Independent Twin Prime Spectrum (ITPS)

The Independent Twin Prime Spectrum is the independent version of $\mathrm{TPS}(x, n)$: all deterministic phases are replaced by an unknown phase $*$ on the same domain.

$$\mathrm{ITPS}(x, n) = w_2\big(x, *_{-1}, 4, n\big) \ \cap \ \bigcap_{i=2}^{s} w_{p_i}\big(x, *_{-2}, p_i^2, n\big).$$

Since $w_2(x, r) = w_2(x, r + 2)$, w_2 loses one phase, while all other waves lose two phases; we denote this by $*_{-2}$.

Now we can calculate the value of the Independent Twin Prime Spectrum.

The formula is computed by prime-square blocks:

$$\big|\mathrm{ITPS}(x, n)\big| = 3 + \sum_{i=2}^{s} \big(p_i^2 - p_{i-1}^2\big) \frac{1}{2} \prod_{j=2}^{i-1} \Big(1 - \frac{2}{p_j}\Big) \ + \ \big(n - p_s^2\big) \frac{1}{2} \prod_{j=2}^{s} \Big(1 - \frac{2}{p_j}\Big)$$

Let us go through an example.

Example ($n = 250$).   Using the actual prime-square blocks up to 250, we obtain:

$$[1, 4): \quad (4 - 1) \cdot 1 = 3$$

$$[4, 9): \quad (9 - 4) \cdot \tfrac{1}{2} = 2.5$$

$$[9, 25): \quad (25 - 9) \cdot \tfrac{1}{2} \cdot \tfrac{1}{3} = 2.66$$

$$[25, 49): \quad (49 - 25) \cdot \tfrac{1}{2} \cdot \tfrac{1}{3} \cdot \tfrac{3}{5} = 2.4$$

$$[49, 121): \quad (121 - 49) \cdot \tfrac{1}{2} \cdot \tfrac{1}{3} \cdot \tfrac{3}{5} \cdot \tfrac{5}{7} = 5.14$$

$$[121, 169): \quad (169 - 121) \cdot \tfrac{1}{2} \cdot \tfrac{1}{3} \cdot \tfrac{3}{5} \cdot \tfrac{5}{7} \cdot \tfrac{9}{11} = 2.8$$

$$[169, 250]: \quad (250 - 169) \cdot \tfrac{1}{2} \cdot \tfrac{1}{3} \cdot \tfrac{3}{5} \cdot \tfrac{5}{7} \cdot \tfrac{9}{11} \cdot \tfrac{15}{17} = 4.17$$

Summing these block contributions gives

$$\big|\mathrm{ITPS}(x, 250)\big| = 22.67 + \epsilon.$$

Let us absorb all remaining discrepancy into ϵ.

Calculating the step sum directly is not computationally pleasant, so we now rewrite the function as a continuous version of the same stepwise sum.

First recall the function

$$G(x, n) := \prod_{x < p \leq n} \Big(1 - \frac{x}{p}\Big).$$

In the special case $x = 2$, this becomes

$$G(2, u) = \prod_{2 < p \leq u} \Big(1 - \frac{2}{p}\Big).$$

Let

$$p_s \leq \sqrt{n} < p_{s+1}.$$

Therefore, the exact finite sum for $|\text{ITPS}(x, n)|$ can be written as

$$|\text{ITPS}(x, n)| = 3 + \frac{1}{2} \sum_{i=2}^{s} \left(p_i^2 - p_{i-1}^2\right) G(2, p_{i-1}) + \frac{1}{2}\left(n - p_s^2\right) G(2, p_s).$$

Since $G(2, u)$ is a step function, this sum is equivalent to the integral form

$$|\text{ITPS}(x, n)| = 3 + \int_2^{\sqrt{n}} u\, G(2, u)\, du.$$

This identity is exact. It is simply the continuous rewriting of the stepwise sum, and it also calculates the effect of parallel waves inside. However, we can see the problem from another side and consider that we have two spectra that are fundamentally independent, meaning they do not have parallel waves. Then their intersection probability should be $G(1, u)^2$, but there are two type of fundamental dependencies, waves $w_2(x, 0)$ and $w_2(x, 2)$ are equivalent also parallel waves do not shape intersection but we already count their value, we need a factor to remove them. That is the reason for the function $4.C_2(n)$.

So

$$G(2, u) = 4\, G(1, u)^2\, C_2(u),$$

where

$$C_2(n) := \prod_{2 < p \leq n} \frac{1 - \frac{2}{p}}{\left(1 - \frac{1}{p}\right)^2} = \prod_{2 < p \leq n} \frac{p(p-2)}{(p-1)^2}.$$

Substituting this into the integral formula gives the exact representation

$$|\text{ITPS}(x, n)| = 3 + 4 \int_2^{\sqrt{n}} u\, G(1, u)^2\, C_2(u)\, du.$$

To estimate this for large n, we use the asymptotic formula

$$G(1, u) \sim \frac{e^{-\gamma}}{\ln u}.$$

Hence

$$G(2, u) = 4\, G(1, u)^2\, C_2(u) \approx 4 e^{-2\gamma}\, \frac{C_2(u)}{(\ln u)^2},$$

and therefore we obtain the finite-n approximation

$$|\text{ITPS}(x, n)| \approx 3 + 4 e^{-2\gamma} \int_2^{\sqrt{n}} \frac{u\, C_2(u)}{(\ln u)^2}\, du.$$

$C_2(n)$ will converge at infinity, and it is known as the twin-prime constant:

$$C_2 := \lim_{u \to \infty} C_2(u) = 0.6601\ldots$$

Then the asymptotic form is

$$\left|\text{ITPS}(x, n)\right| \sim 3 + 4e^{-2\gamma} C_2 \int_2^{\sqrt{n}} \frac{u}{(\ln u)^2} \, du.$$

Finally, after the substitution $t = u^2$, this becomes

$$\left|\text{ITPS}(x, n)\right| \sim 3 + 8e^{-2\gamma} C_2 \int_4^{n} \frac{dt}{(\ln t)^2}.$$

10.2.5 Twin Prime Dependency Factor

We define the *Twin Prime Dependency Factor* by

$$d_{TPS}(n) := \frac{\left|\text{TPS}(x, n)\right|}{\left|\text{ITPS}(x, n)\right|}.$$

We have now formulated both the numerator and the denominator, so the values of $d_{TPS}(n)$ are deterministic and calculable.

Therefore,

$$\left|\text{TPS}(x, n)\right| = d_{TPS}(n) \left|\text{ITPS}(x, n)\right|$$

and we can write the twin-prime counting function as

$$\pi_2(n) = \left|\text{TPS}(x, n)\right| - 3 = d_{TPS}(n) \left|\text{ITPS}(x, n)\right| - 3.$$

10.2.6 Twin Prime Conjecture

Figure 10.4: G. H. Hardy (1877–1947).

The final step in calculating the twin-prime counting function is understanding $d_{TPS}(n)$. We now connect it to the prime dependency factor $d_{PS}(n)$.

Recall that

$$d_{PS}(n) = \frac{|\mathrm{PS}(x, 0, p_i^2, n)|}{|\mathrm{IPS}(x, 0, p_i^2, n)|}.$$

The twin-prime spectrum is the intersection of the two prime spectra at phases 0 and 2. Therefore

$$\mathrm{TPS}(x, n) = \mathrm{PS}(x, 0, p_i^2, n) \cap \mathrm{PS}(x, 2, p_i^2, n).$$

By the Dependency Factor Principle,

$$|A_{\mathrm{dep}}(x, n) \cap B_{\mathrm{dep}}(x, n)| = d_A(n)\, d_B(n)\, |A_{\mathrm{ind}}(x, n) \cap B_{\mathrm{ind}}(x, n)|.$$

Apply this to the two prime spectra. Then

$$|\mathrm{PS}(x, 0, p_i^2, n)| = d_{PS}(n)\, |\mathrm{IPS}(x, 0, p_i^2, n)|,$$

and

$$|\text{PS}(x,2,p_i^2,n)| = d_{PS}(n-2)\,|\text{IPS}(x,2,p_i^2,n)|.$$

Hence

$$|\text{TPS}(x,n)| = d_{PS}(n)\,d_{PS}(n-2)\,\big|\text{IPS}(x,0,p_i^2,n) \cap \text{IPS}(x,2,p_i^2,n)\big|.$$

But

$$\text{IPS}(x,0,p_i^2,n) \cap \text{IPS}(x,2,p_i^2,n) = \text{ITPS}(x,n).$$

Therefore

$$|\text{TPS}(x,n)| = d_{PS}(n)\,d_{PS}(n-2)\,|\text{ITPS}(x,n)|.$$

From the previous chapter we have

$$d_{PS}(n) = \frac{e^{\gamma}}{2}\left(1 + O\!\left(\frac{1}{\ln n}\right)\right).$$

The same form holds for $d_{PS}(n-2)$. Therefore

$$d_{PS}(n)\,d_{PS}(n-2) = \left(\frac{e^{\gamma}}{2}\right)^2\left(1 + O\!\left(\frac{1}{\ln n}\right) + O\!\left(\frac{1}{\ln^2 n}\right)\right).$$

Substituting this into the previous identity gives

$$|\text{TPS}(x,n)| = \left(\frac{e^{\gamma}}{2}\right)^2\left(1 + O\!\left(\frac{1}{\ln n}\right) + O\!\left(\frac{1}{\ln^2 n}\right)\right)|\text{ITPS}(x,n)|.$$

So

$$d_{TPS}(n) = \left(\frac{e^{\gamma}}{2}\right)^2\left(1 + O\!\left(\frac{1}{\ln n}\right) + O\!\left(\frac{1}{\ln^2 n}\right)\right),$$

and in particular

$$d_{TPS}(n) \to \frac{e^{2\gamma}}{4}.$$

Now return to the identity

$$\pi_2(n) = d_{TPS}(n)\,|\text{ITPS}(x,n)| - 3.$$

From the previous section we already obtained

$$|\text{ITPS}(x,n)| \sim 3 + 8e^{-2\gamma}C_2\int_4^n \frac{dt}{(\ln t)^2}.$$

Substituting this into the counting identity gives

$$\pi_2(n) \sim d_{TPS}(n)\left(3 + 8e^{-2\gamma}C_2\int_4^n \frac{dt}{(\ln t)^2}\right) - 3.$$

Now apply the limit

$$d_{TPS}(n) \to \frac{e^{2\gamma}}{4}.$$

Then

$$\pi_2(n) \sim \frac{e^{2\gamma}}{4}\left(3 + 8e^{-2\gamma}C_2 \int_4^n \frac{dt}{(\ln t)^2}\right) - 3.$$

The main integral term simplifies as

$$\frac{e^{2\gamma}}{4} \cdot 8e^{-2\gamma}C_2 = 2C_2.$$

Therefore

$$\pi_2(n) \sim \frac{3e^{2\gamma}}{4} + 2C_2 \int_4^n \frac{dt}{(\ln t)^2} - 3.$$

The constant terms do not affect the asymptotic growth. Hence

$$\boxed{\pi_2(n) \sim 2C_2 \int_4^n \frac{dt}{(\ln t)^2}.}$$

This is the expected main term for the twin-prime counting function.

10.2.7 Twin Prime Effective Spectrum (TPES)

Here $\mathrm{PES}(x, r, 1, n)$ denotes the phase-r effective spectrum on the common endpoint x.
We define the *Twin Prime Effective Spectrum* as

$$\boxed{\mathrm{TPES}(x, n) = \mathrm{PES}(x, 0, 1, n) \cup \mathrm{PES}(x, 2, 1, n)}$$

So TPES can be written as

$$\boxed{\mathrm{TPES}(x, n) = \bigcup_{i=1}^{s} w_{p_i}(x, \{0, 2\}, 1, n),}$$

where $p_1 = 2 < p_2 = 3 < \cdots < p_s$ are the primes with $p_s \leq \sqrt{n}$.

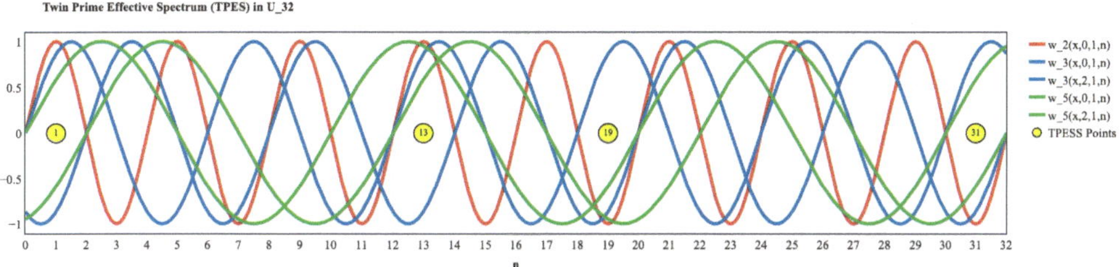

Figure 10.5: TPES(x,32)

Now let us define the shadow of the TPES spectrum.

10.2.8   Twin Prime Effective Shadow Spectrum (TPESS)

Here $\text{PESS}(x, r, 1, n)$ denotes the phase-r effective shadow spectrum on the common endpoint x.
The Twin Prime Effective Shadow Spectrum (TPESS) is

$$\text{TPESS}(x, n) = \text{PESS}(x, 0, 1, n) \cap \text{PESS}(x, 2, 1, n).$$

Equivalently,

$$\text{TPESS}(x, n) = \bigcap_{i=1}^{s} w_{p_i}\big(x,\, R_{p_i} \setminus \{0, 2\},\, 1,\, n\big), \qquad p_s \leq \sqrt{n}.$$

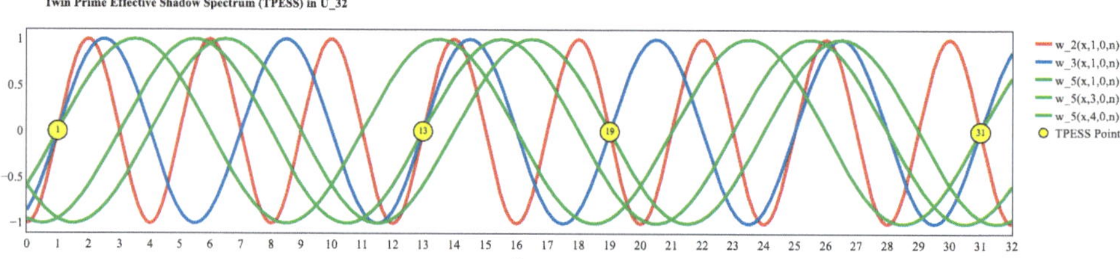

Figure 10.6: TPESS(x,32)

There is a natural relationship between the Twin Prime Spectrum and the Twin Prime Effective
Shadow Spectrum:

$$\text{TPS}(x, n) = \text{TPESS}(x, n) \cup \text{TPS}\big(x, \lfloor \sqrt{n} \rfloor + 2\big).$$

Because twin primes are counted here by their upper endpoint m, the unresolved tail extends to
$m \leq \sqrt{n} + 2$. That is why the exact recursive cutoff is $\lfloor \sqrt{n} \rfloor + 2$.
Also,

$$\text{TPESS}(x, n) \cap \text{TPS}\big(x, \lfloor \sqrt{n} \rfloor + 2\big) = \{1\}.$$

Therefore,

$$\big|\text{TPS}(x, n)\big| = \big|\text{TPESS}(x, n)\big| + \big|\text{TPS}\big(x, \lfloor \sqrt{n} \rfloor + 2\big)\big| - 1.$$

So the twin-prime counting function satisfies

$$\pi_2(n) = \big|\text{TPESS}(x, n)\big| + \pi_2\big(\lfloor \sqrt{n} \rfloor + 2\big) - 1.$$

Thus the problem is reduced to understanding $\big|\text{TPESS}(x, n)\big|$.
We define its dependency factor by

$$d_{\text{TPESS}}(n) := \frac{\big|\text{TPESS}(x, n)\big|}{\big|\text{ITPESS}(x, n)\big|}.$$

Hence

$$\big|\text{TPESS}(x, n)\big| = d_{\text{TPESS}}(n)\, \big|\text{ITPESS}(x, n)\big|.$$

So the remaining task is to formulate the two parts $d_{\text{TPESS}}(n)$ and $\big|\text{ITPESS}(x, n)\big|$.

10.2.9 Independent Twin Prime Effective Shadow Spectrum (ITPESS)

The Independent Twin Prime Effective Shadow Spectrum (ITPESS) is obtained by replacing the deterministic phases 0 and 2 by unknown phases on the same effective domain.

Except for w_2, which loses only one phase, each odd effective prime loses two phases. Therefore

$$\text{ITPESS}(x, n) = w_2\big(x, *_{-1}, 1, n\big) \cap \bigcap_{i=2}^{s} w_{p_i}\big(x, *_{-2}, 1, n\big).$$

So the counting function of the Independent Twin Prime Effective Shadow Spectrum is

$$|\text{ITPESS}(x, n)| = n\left(1 - \frac{1}{2}\right)\prod_{i=2}^{s}\left(1 - \frac{2}{p_i}\right), \qquad p_s \leq \sqrt{n}.$$

Equivalently,

$$|\text{ITPESS}(x, n)| = \frac{n}{2}\, G(2, \sqrt{n}).$$

10.3 The Connection of Twin Primes to the Prime Counting Function

We now connect the twin-prime counting function to the prime counting function.

We begin with the exact identity

$$|\text{TPESS}(x, n)| = \big|\text{PESS}(x, 0, 1, n) \cap \text{PESS}(x, 2, 1, n)\big|.$$

This shows that the Twin Prime Effective Shadow Spectrum is exactly the intersection of two Prime Effective Shadow Spectra, one at phase 0 and the other at phase 2.

By the Dependency Factor Principle, the dependency factor of this moving effective intersection is given by the product of the two corresponding prime-shadow dependency factors, up to the moving-effective error term. Therefore

$$d_{\text{TPESS}}(n) = d_{\text{PESS}}(n - 2)\, d_{\text{PESS}}(n)\left(1 + O\!\left(\frac{1}{\ln n}\right)\right).$$

Hence

$$|\text{TPESS}(x, n)| = d_{\text{PESS}}(n - 2)\, d_{\text{PESS}}(n)\left(1 + O\!\left(\frac{1}{\ln n}\right)\right)|\text{ITPESS}(x, n)|.$$

Now recall that

$$d_{\text{PESS}}(n) = \frac{|\text{PESS}(x, n)|}{|\text{IPESS}(x, n)|} = \frac{\pi(n) - \pi(\sqrt{n}) + 1}{n\, G(1, \sqrt{n})},$$

and also

$$|\text{ITPESS}(x, n)| = \frac{n}{2}\, G(2, \sqrt{n}).$$

Substituting these into the previous identity gives

$$\left|\mathrm{TPESS}(x,n)\right| = d_{\mathrm{PESS}}(n-2)\,d_{\mathrm{PESS}}(n)\left(1 + O\!\left(\frac{1}{\ln n}\right)\right)\frac{n}{2}\,G(2,\sqrt{n}).$$

At this level of asymptotic precision, the shift by 2 does not change the main term, so

$$d_{\mathrm{PESS}}(n-2) = d_{\mathrm{PESS}}(n)\left(1 + O\!\left(\frac{1}{\ln n}\right)\right).$$

Therefore

$$\left|\mathrm{TPESS}(x,n)\right| = d_{\mathrm{PESS}}(n)^2\,\frac{n}{2}\,G(2,\sqrt{n})\left(1 + O\!\left(\frac{1}{\ln n}\right)\right).$$

Now substitute the exact formula for $d_{\mathrm{PESS}}(n)$:

$$\left|\mathrm{TPESS}(x,n)\right| = \left(\frac{\pi(n) - \pi(\sqrt{n}) + 1}{n\,G(1,\sqrt{n})}\right)^2 \frac{n}{2}\,G(2,\sqrt{n})\left(1 + O\!\left(\frac{1}{\ln n}\right)\right).$$

So we obtain

$$\left|\mathrm{TPESS}(x,n)\right| = \frac{\left(\pi(n) - \pi(\sqrt{n}) + 1\right)^2}{2n} \cdot \frac{G(2,\sqrt{n})}{G(1,\sqrt{n})^2}\left(1 + O\!\left(\frac{1}{\ln n}\right)\right).$$

Now use the identity

$$G(2,\sqrt{n}) = 4\,G(1,\sqrt{n})^2\,C_2(\sqrt{n}).$$

Then the factors $G(1,\sqrt{n})^2$ cancel, and we reach

$$\left|\mathrm{TPESS}(x,n)\right| = 2\,C_2(\sqrt{n})\,\frac{\left(\pi(n) - \pi(\sqrt{n}) + 1\right)^2}{n}\left(1 + O\!\left(\frac{1}{\ln n}\right)\right).$$

This gives a direct bridge from the Twin Prime Effective Shadow Spectrum to the prime counting function.

Now return to the exact recursive identity

$$\pi_2(n) = \left|\mathrm{TPESS}(x,n)\right| + \pi_2\!\left(\lfloor\sqrt{n}\rfloor + 2\right) - 1.$$

Substituting the expression for $\left|\mathrm{TPESS}(x,n)\right|$ yields

$$\pi_2(n) = 2\,C_2(\sqrt{n})\,\frac{\left(\pi(n) - \pi(\sqrt{n}) + 1\right)^2}{n}\left(1 + O\!\left(\frac{1}{\ln n}\right)\right) + \pi_2\!\left(\lfloor\sqrt{n}\rfloor + 2\right) - 1.$$

Since

$$\pi_2\!\left(\lfloor\sqrt{n}\rfloor + 2\right) = o\!\left(\frac{n}{(\ln n)^2}\right),$$

the recursive correction is of lower order. Therefore

$$\boxed{\pi_2(n) \sim 2C_2 \frac{n}{(\ln n)^2}.}$$

We can also rewrite the same relation in ratio form. From

$$\pi_2(n) - \pi_2\big(\lfloor \sqrt{n} \rfloor + 2\big) + 1 = |\mathrm{TPESS}(x, n)|,$$

we obtain

$$\boxed{\frac{\pi_2(n) - \pi_2\big(\lfloor \sqrt{n} \rfloor + 2\big) + 1}{\big(\pi(n) - \pi(\sqrt{n}) + 1\big)^2} \approx \frac{2\,C_2(\sqrt{n})}{n} \left(1 + O\!\left(\frac{1}{\ln n}\right)\right).}$$

Finally, using

$$\boxed{C_2(\sqrt{n}) = \prod_{2 < p \le \sqrt{n}} \frac{1 - \frac{2}{p}}{\left(1 - \frac{1}{p}\right)^2},}$$

we get

$$\boxed{\frac{\pi_2(n) - \pi_2\big(\lfloor \sqrt{n} \rfloor + 2\big) + 1}{\big(\pi(n) - \pi(\sqrt{n}) + 1\big)^2} \approx \frac{2}{n} \prod_{2 < p \le \sqrt{n}} \frac{1 - \frac{2}{p}}{\left(1 - \frac{1}{p}\right)^2} \left(1 + O\!\left(\frac{1}{\ln n}\right)\right).}$$

In exact formulas we keep $\lfloor \sqrt{n} \rfloor + 2$, because twin primes are counted by their upper endpoint. In asymptotic formulas this shift contributes only $O(1)$, so it may be suppressed.

Thus, at the asymptotic level, we may write

$$\boxed{\frac{\pi_2(n) - \pi_2(\sqrt{n}) + 1}{\big(\pi(n) - \pi(\sqrt{n}) + 1\big)^2} \approx \frac{2}{n} \prod_{2 < p \le \sqrt{n}} \frac{1 - \frac{2}{p}}{\left(1 - \frac{1}{p}\right)^2} \left(1 + O\!\left(\frac{1}{\ln n}\right)\right).}$$

This is the desired connection: the twin-prime counting function is linked directly to the prime counting function through the finite correction factor $C_2(\sqrt{n})$, or equivalently through its product form.

Exclusive: Prime Patterns Approximation

We have designed Number-Wave Framework and It reveals a lot of mysteries of world of Prime Numbers. We understand and formulate prime numbers. We unlock Twin Primes but it is not the end. With the same strategy we can unlock all the prime patterns. Here is a summary of

$$\gamma \approx 0.5772156649015328606\ldots \qquad e \approx 2.71828182845904523536\ldots$$

$$d_1(n) := \frac{|\mathrm{PS}(x,n)|}{|\mathrm{IPS}(x,n)|}$$

$$d_1 := \lim_{n\to\infty} d_1(n) = \frac{e^{\gamma}}{2} \approx 0.89053620899509899\ldots$$

$$C_2(n) := \prod_{2<p\leq n} \frac{1-\frac{2}{p}}{\left(1-\frac{1}{p}\right)^2}, \qquad C_2 := \prod_{2<p} \frac{1-\frac{2}{p}}{\left(1-\frac{1}{p}\right)^2}, \approx 0.66016181584868695739\ldots$$

$$\pi_2(n) \approx 2\,C_2(\sqrt{n})\,\frac{\pi(n)^2}{n}, \qquad \pi_{\mathrm{cousin}}(n) \approx \pi_2(n),$$

$$\pi_{\mathrm{sexy}}(n) \approx 2\,\pi_2(n), \qquad \pi_{\mathrm{sophie}}(n) \approx \pi_2(n) \cdot \frac{\ln n}{\ln(2n+1)}.$$

$$C_3(n) := \prod_{5\leq p\leq n} \frac{1-\frac{3}{p}}{\left(1-\frac{1}{p}\right)^3}, \qquad C_3 := \prod_{5\leq p} \frac{1-\frac{3}{p}}{\left(1-\frac{1}{p}\right)^3} \approx 0.63516648\ldots$$

$$\pi_{\mathrm{triplet}}(n) \approx \frac{9}{2}\,C_3(\sqrt{n})\,\frac{\pi(n)^3}{n^2}.$$

Here we compare each main counting formula with the exact count at $n = 10^6$. We report the absolute error and the relative error. For all of them, we expect the relative error to converge to zero, while its sign changes infinitely many times. Like strands woven together in a rope, stretching on to infinity.

Table 10.1: Validation at $n = 10^6$ (exact vs. approximations).

Quantity	Exact	Approx	Abs. error	Rel. error
$\pi_2(10^6)$ (twin)	8169	8136.784043	-32.215957	-0.3944%
$\pi_{\mathrm{cousin}}(10^6)$ (gap 4)	8144	8136.784043	-7.215957	-0.0886%
$\pi_{\mathrm{sexy}}(10^6)$ (gap 6)	16386	16273.568085	-112.431915	-0.6861%
$\pi_{\mathrm{sophie}}(10^6)$ $(p, 2p+1)$	7746	7748.051129	+2.051129	+0.0265%
$\pi_{\mathrm{triplet}}(10^6)$ $(p-6, p-4, p)$	1393	1383.061402	-9.938598	-0.7135%

We are now at the threshold of Goldbach.

Dedication

"The most rewarding part is the 'Aha' moment, the excitement of discovery and the joy of understanding something new—the feeling of standing on top of a hill and having a clear view."

— Maryam Mirzakhani

To my mother, **MahBanoo**, who planted in me the belief that no horizon is beyond reach.

To my father, **Amir**, who taught me that perseverance is victory, and that I must never give up.

To my beloved partner, **Dorna**, whose sacrifice, steadfast faith, and enduring support carried me through more than two years of this journey of discovery.

To the great **Mohammad Reza Shah**, whose vision brought universities across Iran and opened the path by which I was able to study, learn, and rise.

And to my hero, **Banoo Maryam Mirzakhani**—a royal soul of mathematics, taken from this world far too soon. She was like a heavenly meteor from the Mother of Science: brilliant, graceful, and unforgettable, illuminating the darkness and revealing hidden gates to all who dare to seek what lies beyond the known.

Maryam Mirzakhani (1977–2017)